Intelligent Systems Reference Library

Volume 164

The aim of this series is to publish a Reference Library, including novel advances and developments in all aspects of Intelligent Systems in an easily accessible and well structured form. The series includes reference works, handbooks, compendia, textbooks, well-structured monographs, dictionaries, and encyclopedias. It contains well integrated knowledge and current information in the field of Intelligent Systems. The series covers the theory, applications, and design methods of Intelligent Systems. Virtually all disciplines such as engineering, computer science, avionics, business, e-commerce, environment, healthcare, physics and life science are included. The list of topics spans all the areas of modern intelligent systems such as: Ambient intelligence, Computational intelligence, Social intelligence, Computational neuroscience, Artificial life, Virtual society, Cognitive systems, DNA and immunity-based systems, e-Learning and teaching, Human-centred computing and Machine ethics, Intelligent control, Intelligent data analysis, Knowledge-based paradigms, Knowledge management, Intelligent agents, Intelligent decision making, Intelligent network security, Interactive entertainment, Learning paradigms, Recommender systems, Robotics and Mechatronics including human-machine teaming, Self-organizing and adaptive systems, Soft computing including Neural systems, Fuzzy systems, Evolutionary computing and the Fusion of these paradigms, Perception and Vision, Web intelligence and Multimedia.

** Indexing: The books of this series are submitted to ISI Web of Science, SCOPUS, DBLP and Springerlink.

More information about this series at http://www.springer.com/series/8578

Laszlo Hunyadi · István Szekrényes
Editors

The Temporal Structure of Multimodal Communication

Theory, Methods and Applications

 Springer

Editors
Laszlo Hunyadi
Department of General and Applied
Linguistics
University of Debrecen
Debrecen, Hungary

István Szekrényes
Institute of Philosophy
University of Debrecen
Debrecen, Hungary

ISSN 1868-4394 ISSN 1868-4408 (electronic)
Intelligent Systems Reference Library
ISBN 978-3-030-22897-2 ISBN 978-3-030-22895-8 (eBook)
https://doi.org/10.1007/978-3-030-22895-8

This Springer imprint is published by the registered company Springer Nature Switzerland AG
The registered company address is: Gewerbestrasse 11, 6330 Cham, Switzerland

This book is dedicated to researchers and developers wishing to deepen their understanding of the complexities of human-human communication as found in multimodal interactions and to seek inspiration to apply specific patterns of behavior in building future smart systems at the crossroads of social sciences and engineering.

Preface

The most widely accepted claim in contemporary multimodality research is that the different channels of communication complement each other. Sharing a similar view and further extending it, this book sets as its main goal to show that the perceived asymmetry between the verbal and nonverbal markers of communication can be reconciled by demonstrating that all these markers are part of a multimodal temporal structure in which their local alignment is just one constituent in a temporally distributed and hierarchically organized structure. By applying a variety of methodologies, including but not restricted to T-pattern analysis, the latter specifically designed for the study of temporal patterns in human interactions, the present volume intends to offer new insights in the theory of multimodal communication as well as pragmatics in general and thus contribute to dialogue modeling both in human and robotic applications.

The most researched challenge in applied multimodal communication is the development of multimodal user interfaces which are intended to make human-computer interaction more human-like and user-friendly. For achieving this goal, multimodal user interfaces (robots or virtual agents) have to be capable of the appropriate interpretation and generation of both verbal and nonverbal behavior (e.g., prosody and gestures) controlled by their communication engine.

Research is highly interdisciplinary involving a number of fields, such as computer and cognitive science, psychology, linguistics, biology, and robotics. The main research areas focus on the following questions:

Motion Intelligence: How can we integrate perception with action to allow robots to move smartly and assist humans in unexpected, everyday environments?

Attentive Systems: What mechanisms enable artificial systems to understand and actively focus on what is important, to ignore irrelevant detail and to share attention with humans?

Situated Communication: How can we coherently coordinate language, perception, and action so that cooperation between humans and artificial systems takes place in natural efficiency?

Memory and Learning: What memory architectures enable a system to acquire, store, and retrieve knowledge, and to improve its capabilities by learning?

There is a wide range of applications from industry to home, from health to security, each having the main goal of uncovering, embedding, and implementing a rich spectrum of knowledge about human behavior and cognition for both understanding ourselves as humans in general, and the building of more effective human-machine interfaces. This volume offers a broad view on the theory, methodology, and practice of the study of multimodal behavior that can be applied to a number of future applications. The authors are representatives of fields of research that can significantly contribute to the achievement of the above goals, coming from specific areas of psychology, computer science, speech science, pragmatics and computational linguistics. The topics presented in the book include an overview of the theory of pragmatics related to multimodal communication, a methodology for the detection of structures in indirect observation, an experiment on how the involvement of a direct observer in a dialogue can be measured, a description of a methodology of discovering and applying paraverbal, kinetic behavioral patterns in an educational environment, the description of multimodal patterns of uncertainty in a dialogue, and how machine learning techniques can be applied to the automatic recognition of topic change in a conversation. Time plays an important role throughout the book, demonstrating that behavior needs to be captured as a function of time across all modalities.

Special features of the book:

1. It offers a comprehensive analysis of multimodal communication based on a variety of corpora, including one of the most extensively annotated corpora of dialogues (HuComTech).
2. It captures the temporal patterns of a wide range of behaviors discovered in human-human communication.
3. It presents work from several areas of communication research with all the benefits of interdisciplinarity.

Debrecen, Hungary   Laszlo Hunyadi
January 2019   István Szekrényes

Contents

Contributors

M. Teresa Anguera Faculty of Psychology, Institute of Neurosciences, University of Barcelona, Barcelona, Spain

Oleguer Camerino National Institute of Physical Education of Catalonia (INEFC), University of Lleida, Lleida, Spain;
Lleida Institute for Biomedical Research Dr. Pifarré Foundation (IRBLLEIDA), University of Lleida, Lleida, Spain

Nick Campbell Speech Communication Lab, Trinity College Dublin, Dublin 2, Ireland

Marta Castañer National Institute of Physical Education of Catalonia (INEFC), University of Lleida, Lleida, Spain;
Lleida Institute for Biomedical Research Dr. Pifarré Foundation (IRBLLEIDA), University of Lleida, Lleida, Spain

Anna Cuxart Universitat Pompeu Fabra (retired professor), Barcelona, Spain

Ghazaleh Esfandiari-Baiat Department of General and Applied Linguistics, University of Debrecen, Debrecen, Hungary

Anna Esposito Department of Psychology, Seconda Universitá di Napoli, Caserta, Italy

Robin M. Hogarth Department of Economics and Business, Universitat Pompeu Fabra, Barcelona, Spain

Laszlo Hunyadi Department of General and Applied Linguistics, University of Debrecen, Debrecen, Hungary

György Kovács Research Institute for Linguistics of the Hungarian Academy of Sciences, Budapest, Hungary;
MTA SZTE Research Group on Artificial Intelligence, Szeged, Hungary;
Embedded Internet Systems Lab, Luleå University of Technology, Luleå, Sweden

Enikő Németh T. Department of General Linguistics, University of Szeged, Szeged, Hungary

Mariona Portell Universitat Autònoma de Barcelona, Cerdanyola del Vallés, Spain

István Szekrényes Institute of Philosophy, University of Debrecen, Debrecen, Hungary

Part I
Theoretical Overview of Multimodal Communication

Chapter 1
Linguistic and Contextual Clues of Intentions and Perspectives in Human Communication

Enikő Németh T.

Abstract Social forms of language use such as communication, information transmission without communicative intention, and manipulation may be distinguished on the basis of language users' intentions. Language users' perspectives get tied up with their intentions. Identifying intentions in language use is not an easy task. Usually language users provide various clues to make it possible for their partners to take their perspectives and infer their intentions. The present chapter seeks to investigate these clues, concentrating on the role of some linguistic devices and contextual clues. Before discussing intentionality and perspectivity, the chapter overviews the main types of human communication models and examine whether they are suitable for handling the multimodality, temporal sequentiality, perspectivity, and intentionality of human communication. Then, following a thorough analysis of some data from a thought experiment and a corpus, first, the chapter analyzes how speakers can realize their intentions and what perspectives speakers attempt to develop in their partners in order to achieve that their partners infer the intentions the speakers wish to convey. Second, it examines how partners can infer speakers' intentions based on the clues provided by the speakers. And, third, it extends the analysis to information transmission without communicative intention and manipulation.

1.1 Introduction

The study of language use mainly focuses on an examination of verbal communication, but it is not the only form of language use. There are various other forms including the use of language to inform, manipulate, think, memorize, learn, play a game, and sing for fun (Németh T. 2008, 153–154, 2015, 53). Various forms of

The research results of which are presented here has been supported by the MTA-DE-SZTE Research Group for Theoretical Linguistics as well as by the EU-funded Hungarian grant EFOP-3.6.1-16-2016-00008.

E. Németh T. (✉)
Department of General Linguistics, University of Szeged, Szeged, Hungary
e-mail: nemethen@hung.u-szeged.hu

© Springer Nature Switzerland AG 2020
L. Hunyadi and I. Szekrényes (eds.), *The Temporal Structure of Multimodal Communication*, Intelligent Systems Reference Library 164,
https://doi.org/10.1007/978-3-030-22895-8_1

language use can be distinguished according to their social attributes. In a social form speakers make utterances toward other persons, while in individual forms such as using language to think, memorize, learn, and sing for fun while taking a bath, utterances are not addressed to other persons. Social and individual forms of language use may be culturally different (Németh T. 2014, 2015).

Social forms of language use such as communication, information transmission without communicative intention, and manipulation can be distinguished on the basis of language user's intentions (Németh T. 2008, 2014, 2015). In social ways of language use, language users take their partners' perspectives and change their initial egocentric perspectives (Bezuidenhout 2013). Language user perspectives contain their intentions; hence, it is reasonable to assume an intentional viewpoint within perspectives in addition to the spatial, social etc. viewpoints (Németh T. 2014, 2015). In communicative interactions the speakers' and hearers' intentional viewpoints coincide to a great extent; the speakers attempt to realize their informative and communicative intentions and the hearers recognize them (Sperber and Wilson 1986/1995). However, in informative and manipulative forms of language use the speakers' and hearers' intentional viewpoints differ from each other, since the speakers do not want to reveal all the intentions they have and the hearers do not recognize all the intentions the speakers have. The hearers are supposed to recognize only the intentions the speakers want to be recognized within a perspective the speakers want the hearers to take. Based on the extent to which language users' intentional viewpoints coincide or differ from each other, we may be able to make certain predictions regarding the success of communicative, informative and manipulative forms of language use.

Identifying intentions in social forms of language use is a difficult task. Therefore language users provide various clues to enable their partners to take their perspectives and infer their intentions. In the present chapter I will investigate these clues concentrating especially on the role of some linguistic devices and contextual clues including general pragmatic knowledge, cultural knowledge and particular contextual information in verbal communication. In accordance with the general basic assumptions of this volume, I examine human communicative interactions in which communicative partners apply a verbal code inherently multimodal and temporally sequential in each modality. Taking into consideration this multimodality and temporality and relying on a thorough analysis of some data from a thought experiment and a corpus, first, I seek to analyze how speakers can realize their intentions through their perspectives in communication and what perspectives speakers attempt to develop in their partners in order to achieve that their partners infer the intentions the speakers want to reveal. Second, I seek to examine how partners can infer speakers' intentions taking their perspectives on the basis of their own perspectives and the clues provided by the speakers. Afterwards, I will mention that the analytical method used in this chapter can be applied to the other social forms of language use, namely information transmission without communicative intention and manipulation, and its results can be utilized in the development of the models of human-computer interaction.

The organization of the chapter is as follows. In Sect. 1.2, I will discuss the main types of models of human verbal communication and examine whether they are suitable for handling the multimodality and temporal sequentiality of human verbal communicative interactions based on the perspectives and intentions of communicative partners. In Sect. 1.3, I will investigate the role of perspectives and intentions in verbal communication in detail, concentrating on their verbal and contextual clues. Lastly, in Sect. 1.4, I will summarize my findings.

1.2 Models of Human Verbal Communication

1.2.1 The Basic Features of Human Verbal Communication

When examining communication models, we would like to determine what basic features can be assumed in human communicative interactions which have to be accounted for in a model which seeks to explain communication. In human communication, communicators use all possible means available for them to convey pieces of information explicitly or implicitly that they wish to be processed by their communicative partners. Taking into account their partners' perspectives, communicators intentionally produce physical stimuli based either on verbal and non-verbal code-systems or situation-bound behavior without code using to transfer the intended mental representations meant for communicative partners (Németh T. 2015). Communicative partners are supposed to process all physical stimuli available from various sources in communicative interaction to construct mental representations. In the course of information processing, taking into consideration the communicators' perspectives and intentions, the communicative partners rely on the verbal and non-verbal code-systems as well as situation bound behavior without code using applied by the communicators to grasp the explicitly conveyed pieces of information, and, in addition, they consider the context to get the implicitly induced levels of information (Németh T. 2011, 43). Thus, verbal communication may be characterized as a dynamic interaction that not only consists of the production and interpretation of utterances as verbal stimuli, but also involves non-verbal signs such as acoustic and visual stimuli, and situation-bound behavior without code using (Németh T. 2011, 44). Consequently, verbal communication may be viewed as inherently multimodal. In the flow of communication, a pragmatic division of labor between the operations of separate modalities may be observed (Hunyadi 2011; Németh T. 2011, 50–52). Let us consider the English adaptation of Németh T.'s (2011, 51) example devised in a thought experiment.

(1)   *B* is working on her laptop. She is engaged in writing a scientific paper. *A* enters the room, goes to *B*, puts his hand on *B*'s shoulders, and whistles for her. B turns to A and they make eye contact with each other. Then *A* says. – Let's go to the movies tonight evening. *B* raises her eyebrows, directs her gaze to the laptop, points to the laptop, and says. – Tomorrow is the deadline.

The interaction in (1) may be interpreted as a manifestation of multimodal communication. The participants *A* and *B* produce various physical stimuli belonging to different modalities. For instance, when *A* puts his hand on *B*'s shoulders, he uses a conventional non-verbal sign of attracting one's attention. At the same time, whistling by which *A* also wants to attract *B*'s attention is an ostensive behavior[1] without code using since in other situations whistling does not necessarily serve as a behavior of attracting one's attention. For example, one can whistle for fun only. Furthermore, making eye contact is also a conventional, non-verbal sign to initiate a communicative interaction and the production of the utterance *Let's go to the movies tonight evening* is a manifestation of verbal code using, more precisely, it is an explicitly conveyed direct illocutionary act. *B*'s communicative act in (1) can be analyzed in a similar way. Raising the eyebrows is a conventional non-verbal sign by means of which the agent expresses her/his surprise. The directing of the gaze to the laptop and pointing to the laptop are kinds of ostensive behavior without code using, while the production of the answer *Tomorrow is the deadline* is a verbal communicative indirect act.

The analysis of *A*'s and *B*'s multimodal communicative interaction in (1) demonstrates that the use of non-verbal and verbal codes as well as situation bound ostensive behavior without any code using express the intended information together in an interaction that strengthens each other's contribution to the information transmission. Consequently, the various modalities applied by *A* and *B* in (1) are not independent of each other in the sense that they all contribute to the meaning construction and interpretation in the communicative interaction. From another point of view, these modalities may be considered independent of each other since each of them can convey information separately. For example, if *B* does not say a word in (1) and only raises her eyebrows or directs her gaze to the laptop or points to the laptop, she can also express her indirect refusal to go to the movies with *A*. At the same time, it may also happen that the pieces of information originated in various modalities in communication are contradictory. In this case, non-verbal communication and communicative behavior without code using can override the information transferred by verbal communication (Buda 1988; Németh T. 1990).[2]

In addition to perspectivity, intentionality and multimodality discussed above, there is a fourth essential feature of communication, namely temporal sequentiality (Németh T. 2011, 53). Communication is a dynamic process which consists not only of the production and interpretation of one utterance, but also contains sequences of multimodal communicative acts, and the roles of actual speakers continuously change between the participants. The explanation of the dynamic sequential organization of communication is necessary for the description of the contributions of participants to the overall coherence of discourses in communication (Goodwin 1981; Abuczki 2011).

---

[1]Ostensive behavior is a kind of behavior by means of which the agent makes manifest his/her intention to make something manifest (Sperber and Wilson 1986/1995, 49).

[2]Overriding the information provided by the verbal code can be explained by the assumption that the use of a verbal code (i.e. natural language use) entered the communication in a later phase from an evolutionary point of view (Németh T. 2005).

If we treat communication as a sequence of communicative acts then in the course of sequential analysis we have to take into account not only the verbal modality but non-verbal ones as well (Abuczki et al. 2011). On the basis of the quantitative and qualitative analyses of the dialogues in the pragmatically annotated HuComTech corpus,[3] (Abuczki 2011) demonstrated that the elements of the verbal and non-verbal codes, i.e. acoustic and visual signs are related to each other in the organization of communicative events, or in other words, the sequential organization of various modalities together determine the dynamic flow of the interaction (cf. also Ford and Thompson 1996; Haugh 2010).

However, one can ask whether situation bound communicative behavior without code using has a role in the sequential organization of communicative interactions. (Abuczki 2011) did not examine this kind of communication in the HuComTech corpus; instead, she focused on the interaction between the signs of verbal and non-verbal codes in organizing communicative interaction with a special attention to turn sequences.[4] But if we return to the communicative interaction in (1) we may realize that the whistling, the gazing and pointing to the laptop are various kinds of situation-bound ostensive behavior without code using that not only strengthen the information arising from the use of physical stimuli of verbal and non-verbal codes, but they also signal the start of the communicative interaction and turn initiation, i.e. they also play a role in the sequential construction of communicative interactions. Hence, it is plausible to assume that the sequential organization of communicative interactions also have a multimodal nature.

In summary: on the basis of the characterization of communication provided in the previous paragraphs, the expression of the communicative partners' intentions and perspectives as well as sequentiality in communication can only be described and explained by taking into account multimodality. Consequently, an adequate communicative model has to cover all these aspects of communicative interactions. In the following Sects. 2.2–2.5, I will examine whether the most popular communication model types such as code model, inferential model, ostensive-inferential model and generative model can handle these aspects of communication.

1.2.2 Code Model

The models of human communication can be divided into different groups according to what aspects of communicative interaction they concentrate on and how

---

[3]The HuComTech corpus is a multimodal Hungarian spoken language corpus consisting of about 50 hours of video and audio recordings of 111 formal dialogues (simulated job interviews) and 111 informal but guided dialogues. The corpus was built at the Department of General and Applied Linguistics, University of Debrecen under the supervision of László Hunyadi. The participants were university students. The corpus has both a unimodal and a multimodal annotation. For a detailed description of the corpus see https://tla.nytud.hu/MTARIL/01-0003-0000-0000-0000-1@view.

[4]Sometimes it is not easy to decide whether a piece of communicative behavior is a manifestation of the use of a non-verbal code system or it is a kind of ostensive behavior without any code using.

they attempt to describe and explain communication. Starting from the Aristotelean approach, code models in various disciplines such as e.g. cybernetics, mathematics, informatics, linguistic communication studies, and pragmatics consider communication a process between two information processing devices. In the course of communication one of these information processing devices modifies the other's physical environment in order to attract the other's attention and to transfer the intended pieces of information by means of the produced physical stimuli which is an element of a verbal or non-verbal code. The other information processing device realizes that something has been modified in the physical environment, directs his/her attention to the physical stimulus produced by his/her partner and processes the transferred information using the same code (Shannon and Weaver 1949; Toda 1967; Jakobson 1960; Németh T. 2005). The aim of the interaction between the two participants is that both of them have the same mental representation. The condition for having the same representation is the knowledge and use of the same code system in communication. According to code models, communication is a coding-decoding procedure of explicitly conveyed messages. The success of communication is guaranteed when communicative partners share the knowledge of the code applied in the interaction.

Although code models have a long tradition, there are several problems with them (Sperber and Wilson 1986/1995, 1–15, Németh T. 1996, 10–11, 2011, 47–48). First, they are descriptively inadequate since they only focus on the kind of communication which is based on code using. They cannot handle situation bound ostensive behavior without code using. Second, since the communication is only based on the coding-decoding of messages built from the elements of a code, code models cannot take into account the intentions and perspectives of participants which are not expressed explicitly. Third, following from this, code models cannot describe the context-dependent non-literal meanings such as e.g. irony, live metaphors, indirect speech acts, and implicatures since there is no place for inferences in them. Fourth, code-models are not suitable for providing a description of multimodality and sequentiality in communication. Although they can explain the information transmission based on different verbal and non-verbal codes separately, they cannot take into consideration their interaction with each other and situation bound ostensive behavior without code using, neither in the meaning construction of multimodal communicative acts nor in multimodal sequential organization in discourses (cf. the analysis of (1) in Sect. 1.2.1).

1.2.3 Inferential Model

According to the inferential models of communication, the essence of communication is not the coding and decoding process of explicit messages but the expressing and realizing of intentions of participants, i.e. the conveying and inferring of the speaker's meaning (Grice 1957, 1975; Lewis 1969). Communication is successful when the communicative partner can interpret the meaning of utterances produced by the communicator in the particular contexts. The speaker's meaning intended by the communicator usually is richer than the explicitly expressed linguistic meaning of the sentences uttered, i.e. the coded meaning of utterances is underdetermined (Horn 2004; Ariel 2008). By obeying the Cooperative Principle and flouting its maxims, the communicator can implicate items of information in addition to what he/she explicitly expresses with the uttered linguistic forms. The speaker's meaning can be divided into two parts: what is said and what is implicated (Grice 1975); i.e. the linguistic coded meaning and implicated meaning together provide the meaning intended by the speaker when communicating. The communicative partner also relies on the Cooperative Principle and its maxims in the course of the construction of the meaning of utterances intended by the communicator (Grice 1975). The communicative partner decodes the explicitly expressed linguistic meaning and infers the implicated meaning.

Although the inferential models have more descriptive power than the code models, they are also inadequate to describe communication. Inferential models focus mainly on the inferential side of communication, i.e. the communicative partner's side. Therefore inferential models are suitable for explaining how a hearer can infer implicatures within the speaker's meaning in the course of the interpretation of individual utterances. At the same time, they do not concentrate on the communicator's side. Furthermore, inferential models neither examine the relation between what is said and what is implicated nor the relation between the communicator's and the partner's sides (Németh T. 2011, 48). Consequently, inferential models may be considered one-sided and not dynamic, and following from this they probably cannot handle the dynamic sequentiality of communication. What is more, since they focus mainly on the inferring of implicatures, they do not sufficiently take into consideration the interaction between verbal and non-verbal code using as well as situation bound ostensive behavior without code using, i.e. the multimodality of communication cannot be described in inferential models in an adequate way.

1.2.4 Ostensive-Inferential Model

The ostensive-inferential model of communication constructed by Sperber and Wilson (1986/1995) defines communication as follows:

(2)   "the communicator produces a stimulus which makes it mutually manifest to communicator and audience that the communicator intends, by means of this stimulus, to make manifest or more manifest to the audience a set of assumption $\{I\}$" (Sperber and Wilson 1986/1995, 63)

In this definition two intentions are hidden, namely the informative and communicative ones. The content of the informative intention is to inform the communicative partner about a set of assumptions $\{I\}$, while the content of communicative intention is to make mutually manifest the communicator's informative intention (Sperber and Wilson 1986/1995, 58, 61). It is obvious that in Sperber and Wilson's opinion

(1986/1995, 64) communication is always intentional for at least two reasons. The first reason is a Gricean one: by producing direct evidence of one's informative intention, one can convey a wider range of information than by conveying direct evidence only for the content of information. And the second reason that humans have for communicating is to modify and extend each other's cognitive environment. Although Sperber and Wilson consider ostensive-inferential communication intentional, they also emphasize that their definition does not exclude the possibility of unintentional communication (Sperber and Wilson 1986/1995, 63–64). A stimulus merely intended to convey information might make mutually manifest the informative intention without a communicative intention, and this, in accordance with Sperber and Wilson's definition of ostensive-inferential communication, would count as communication (Németh T. 2014, 474, 2015, 56). Sperber and Wilson (1986/1995, 64) call this form of information conveying a case of unintended, covert communication. However, on the basis of the conception of ostension as well as informative and communicative intentions provided by Sperber and Wilson (1986/1995, 49–63), intentionality should be a defining feature of communication, therefore the definition of communication in (2) requires some modification (Németh T. 2014, 474, 2015, 57):

(3)   "the communicator produces a stimulus by which the communicator makes it
        mutually manifest to herself/himself and the audience that the communicator
        intends, by means of this stimulus, to make manifest or more manifest to the
        audience a set of assumptions $\{I\}$"

The modified model of ostensive-inferential communication in (3) can handle the intentionality of communication and eliminate some disadvantages of code models and inferential models. For instance, in the case of verbal communication, the explicit linguistic stimulus is produced by the communicator by coding and interpreted by the communicative partners by decoding, while implicit meanings are produced by ostension without coding and interpreted by inferring (cf. Németh T. 2008).

As we have seen above, human verbal communication can be characterized as inherently multimodal,[5] since it involves the use of verbal codes, several signs of non-verbal codes and ostensive behavior without code-using as well (cf. Ivaskó and Németh T. 2002; Németh T. 1990, 2005, 2008). Sperber and Wilson's original model of ostensive-inferential communication and its modified version suggested by Németh T. (2014, 2015) do not describe this multimodality, so they must be extended using it. A communicative act may contain stimuli from more than one modality. Various functions can be assigned to these joint occurrences of the different modalities. The jointly occurring modalities can not only strengthen, override or complement each other as was supposed in the literature previously, but they can

---

[5]It is worth mentioning that non-verbal communication can also be multimodal since codes from more than one non-verbal modality can be used in it. Furthermore, situation-bound ostensive behavior without code using can also be multimodal performing such behavior which is based on visual, auditory etc. stimuli.

also strongly interact in multiple ways in constructing human interaction, i.e. they co-construct multimodal human communication.

Besides the lack of involving multimodality into the definition, the ostensive inferential communication model in (2) and its modified version in (3) have another inadequacy: they do not take into consideration the sequential structure of communication. Communication is a two (or more)-sided dynamic process which consists not only of the production and interpretation of a unique stimulus, but also involves sequences of multimodal communicative acts constructing coherent discourses. Formal, computational approaches to communication analyze coherent discourses and mainly rely on two different traditions (Bunt and Black 2000; Jurafsky 2004). The first tradition starts out from the communicators' intentions, beliefs and desires in accordance with classical philosophical inferential pragmatics that focuses not only on the hearer's side of communication but also the production side. The second tradition relies on the surface clues of discourses in accordance with code models. In these computational models multimodality has not been taken into account.

After considering the above argumentation for the integration of intentionality, perspectivity, sequentiality, and multimodality into a dynamic model of communication, let us modify the definition in (3) as follows.

(4)   From a shared perspective, the communicator produces stimuli in various modalities by which the communicator intends to make it mutually manifest to herself/himself and the audience that the communicator intends, by means of these stimuli, to make manifest or more manifest to the audience a set of assumptions {I}. Interpreting the stimuli in various modalities produced by the communicator, the audience should realize the communicator's intentions and process the set of assumptions {I} in their shared perspective.

Five remarks seem to be in order for the definition in (4). First, the plural form of the term *stimulus* makes it possible to incorporate sequentiality into the definition. It means that communication consists not only of one stimulus, but also a sequence of stimuli. Second, the plural form *stimuli* can also mean that the communicator produces more than one stimulus at the same time but in different modalities. Third, the definition explicitly says that a shared perspective has to be assumed for the participants in the communicative interaction.[6] Fourth, the definition takes into account both the production and the interpretation sides of communication. And fifth, the definition allows for the roles of participants to be modified, i.e. communicative partners may become communicators when they produce stimuli and communicators can become communicative partners when they interpret the communicators' stimuli.

---

[6]For a more detailed discussion of perspectivity in communication see Németh T. (2014, 2015) as well as Sect. 1.3

1.2.5   The Generative Theoretical-Technological Model

Hunyadi (2011, 2013) suggests a formal, generative theoretical-technological model for human communication with an emphasis on the requirements of the possibility of the implementation for a human-computer interaction. According to his model, a communicative event can be characterized as multimodal, modular, and generative (Hunyadi 2013, 187). Multimodality in Hunyadi's definition means that an event is a complex of verbal and nonverbal components even in the case where it is mainly based on the use of a verbal code (Hunyadi 2013, 187). However, it should be added that in Hunyadi's (2011, 2013) model multimodality only refers to the part of communication which is primarily based on verbal and non-verbal code using. It does not focus on situation-bound ostensive behavior without code using, so a key part of communication has not been covered by this model.

The second criterion for describing a communicative event is modularity (Hunyadi 2013, 187). Hunyadi assumes that functions and meanings in multimodal communication are realized in various modules which are not independent of each other in the sense that the output of a module can serve as an input of another module, or in other words, the output of a module feeds the input of another module. Three main modules were proposed by Hunyadi (2011, 2013). The first module is responsible for the underlying general frame of all possible communicative events at a purely formal level. At this level the formal basic structure of communication including the sequentiality can be grasped. The output, i.e. the formal basic structure of communicative events serves as an input for the second module. The second module provides a finite set of non-contextualized communicative functions at the level of functional extension. And finally, the third module assigns context-dependent particular communicative events to the functionally extended basic structures at the level of pragmatic extensions. The communicators' particular intentions and perspectives can be described at this level. Underlying this view of the modularity of communication there is a rule-based approach according to which communication has an underlying formal structure with certain possible functional extensions which are realized as particular context-dependent communicative events at the surface.

The third important feature of Hunyadi's (2011, 2013) model of multimodal communication is generativity. Minimal ingredients are assumed at every level called primitives, from which all and only those communicative events can be generated which are evaluated as acceptable on the basis of the language users' intuition within a particular context. These primitives are mapped onto specific markers in each modality and these modality specific markers may be observed in a particular communicative interaction.

Contrary to the intention- and clue based computational approaches, Hunyadi's (2011, 2013) model attempts to take into account both sides and directions of communication, i.e. the production (synthesis) as well as the interpretation (analysis). In the course of the modeling of human grammatical and pragmatic competences, generative linguistics starts out from an underlying structure. According to Hunyadi's model, communication also has a bare underlying structure without semantics. The basic idea is that when a communicator communicates with his/her partner, he/she takes into consideration not only his/her own intentions/beliefs/desires but also the general underlying structure of communication, the potential communicative procedures, context and the multimodal tools (clues) which can be applied in order to achieve his/her communicative goals. And vice versa: when a communicative partner tries to interpret the communicator's multimodal communicative behavior he/she relies not only on the surface clues available in the discourses but also his/her knowledge about the underlying structure of communication, context as well as his/her beliefs about the communicator's intentions/beliefs/desires. Consequently, Hunyadi's model can combine the advantages of the above-mentioned two rivaling computational pragmatic approaches. However, Hunyadi's model also requires some modification, because it takes into consideration only code-guided behavior, i.e. it describes only verbal and nonverbal modalities based on a natural language use or non-verbal code-using. Hunyadi's rule-based approach must be extended with the production and interpretation of ostensive behavior without code-using.

After discussing the main types of communication models and asking whether they can handle intentionality, perspectivity, sequentiality, and multimodality of communication, let us turn to the problems of the manifestation of communicative partners' intentions and perspectives in communicative interactions.

1.3 Perspectives, Intentions and Their Clues in Verbal Communication

In two previous papers, I defined perspective as an initially egocentric complex mental position of a language user which is grounded in her/his individual brain. In social forms of language use speakers and listeners should take into account their partner's perspective from the very beginning of the interaction altering their own initially egocentric perspective (Németh T. 2014, 2015). This is the phenomenon of perspective-taking which can be interpreted in linguistic utterances and other accompanying ostensive stimuli performed by actual speakers (Sanders and Spooren 1997, 89–91; Tomasello 1999; MacWhinney 2005). However, it is worth recalling that people cannot usually set aside entirely their own perspective when they take another's, because perspective-taking can also occur only through an initial, individual egocentric one which is affected by taking another person's perspective.

Perspective, as a complex mental position, is basically formulated from two kinds of information in contexts of language use. The first information package contains all perceivable pieces of information. These are the elements of the directly observable physical and social environment, such as time, space, different kinds of participants, and social relations. According to the observable categories, spatial, temporal, and social viewpoints may be assumed within one's perspective. The other kind of information in one's perspective includes her/his mental states, such as representations of experiences, background knowledge, emotions, and attitudes. The social viewpoint is also a complex position, which is formed not only from information originating from perceptions, but also from general and particular background knowledge concerning, for instance, relatives' relations or the organization of a society, which may be manifested, for instance, in culturally different honorific systems.

To have intentions is possible within a particular perspective and to infer someone's intentions is only possible if we take her/his perspective. Thus, as I mentioned earlier, it is reasonable to assume an intentional viewpoint within one's perspective similar to temporal, spatial, and social viewpoints. The perspectival nature of intentions can also be detected in distinguishing between the social forms of language use.

The different viewpoints in one's perspective have various linguistic and contextual indicators. Linguistic indicators of temporal, spatial, and social viewpoints are usually placed under the label of deixis. Traditional categories of deixis such as person, time and place as well as additional categories such as social deixis (cf. various kinds of honorifics) and discourse deixis may be organized in an egocentric way (Levinson 1983, 54–68) or in a way which mirrors the other persons' perspectives. Let us take Levinson's example for person deixis, in which a wife utters (5) to her husband in the presence of their son, little Billie:

(5)   Can Billie have an ice-cream, Daddy?

The wife communicates with her husband: her utterance in (5) is addressed to her husband, who is Billie's father and since Billie is also present in the speech situation, the wife takes Billie's social point of view and uses *Daddy* for the purpose of vocative selection. The perspective taking is mirrored in the category of person deixis. If the wife behaves only egocentrically based on her own social viewpoint, she should say something like (6):

(6)   Can Billie have an ice-cream, Honey?

In (6) the utterance is addressed to the husband from the wife's perspective using *Honey* for the purpose of vocative selection.

Let us conduct a thought experiment and imagine a situation in which the father has punished the three year old little Billie for his naughty behavior and prohibited him eating ice-cream. In such a situation the father starts eating ice-cream and little Billie, who would like to eat it as well, does not want to ask his father to let him eat ice-cream. The mother who does not know anything about the punishment enters the room and sees that the father is eating ice-cream and little Billie is watching the father ravenously. After a while little Billie glances at his mother who realizes her son's asking gaze. On the basis of her visual perception of the situation and the ostensive stimulus, i.e. asking gaze produced by his son as well as her previous background knowledge that little Billie likes ice-cream very much, the mother takes the perspective which she thinks it is Billie's and according to which she makes an indirect request to the father on behalf of little Billy to give him ice-cream. In other words, the wife asks her husband as Billie would ask, i.e. from Billie's perspective but in her own name. The indirect request *Can Billie have an ice-cream, Daddy?* in the wife's utterance may be interpreted in another way as well, namely as an echoic one. It is quite usual that little children refer to themselves by their first name instead of using personal pronoun *I*. Taking into account this kind of person deixis applied by little children, the wife makes her son's request manifest to the father by uttering this on behalf of her son, i.e. the mother echoes her son's unarticulated utterance.

The wife's perspective taking is induced by several factors: (i) her empirical observation of the particular situational context, (ii) Billie's ostensive stimulus (asking by gaze), (iii) her background knowledge about her son's habits and favorite sweets, and, (iv) her intention to help Billie. These factors motivate the mother to assume that Billie has an intention to ask his father to give him an ice-cream, but he does not dare to perform the request for some reason.

Let us notice that both the mother's initial perspective and perspective which she takes believing that it is her son's include various intentions. For example, the mother may infer from the above-mentioned pieces of information that Billie would like to eat ice-cream and since eating it in this situation is only possible if Billie performs the intentional communicative act of requesting toward his father and because according to the mother's presumption Billie does not dare and/or does not want to ask, the mother herself performs the conventional indirect request in (5) on behalf of Billie. A conventional indirect request by itself is a clue to the user's intentions since it has language specific linguistic elements which conventionally indicate the speaker's intention to communicate, in other words, the speech act of requesting is inherently a communicative act. Thus, in (5) the wife has two intentions to communicate and request in communication. The conventional linguistic tools of indirect requesting and asking gaze are ostensive stimuli involved in the general pragmatic knowledge concerning the speech act of requesting. The vocative form *Daddy* can also be considered a linguistic tool of indicating the speaker's social viewpoint in her/his perspective and it can determine the speaker's intentional viewpoint indicating that the speaker has a communicative intention toward the hearer. Hence, the use of *Daddy* mirrors the mother's intention to communicate and intention to reveal that she is speaking on behalf of her son.

In (6) the wife also performs an indirect request addressed to the father, but not from her son's perspective. It is involved in the general pragmatic knowledge of American English native speakers that the members of a couple very often turn to each other by means of the vocative form Honey which is a culturally determined conventionalized tool to address utterances to the other partner. Therefore, the vocative form *Honey* indicates that the mother turns to the father as a wife to her husband and it also predicts that the wife's utterance will be interpreted in her perspective. With the use of *Honey*, the wife does not have any intention to request her husband from her son's perspective or to echo her son's unarticulated indirect request addressed to the father. She turns to her husband from her own perspective. By using *Honey* it is not possible for the mother to request the father to give ice-cream to Billie on behalf of Billie; it is only possible to request the husband—and not the father—instead of Billie.

Let us take a third version of the request and replace *Honey* by *Darling*.

(7)   Can Billie have an ice-cream, Darling?

One can ask whether there is any difference between the utterances with *Honey* and *Darling*.[7] In the sense that both utterances can be interpreted as a request, there is none. However, since in various English speaking cultures there may be a difference between the use of *Honey* and *Darling*, the two utterances can be interpreted differently.

For instance, the Collins COBUILD English dictionary for advanced learners (Sinclair 2001, 752) says that the vocative form Honey is American English usage:

(8)   1.Honey is a sweet, sticky, yellowish substance that is made by bees.
        2.You call someone **honey** as a sign of affection. [mainly AM]
          Honey, I don't really think that's a good idea.

The word darling is defined in COBUILD dictionary (Sinclair 2001, 380–381) as follows.

(9)   1.You call someone **darling** if you love them or like them very much.
          Thank you, darling.Oh, darling, I love you.
        2.In some parts of Britain, people call other people **darling** as a sign of
          friendliness.
        3.Some people use **darling** to describe someone or something that they love
          or like very much. [INFORMAL]
          To have a darling baby boy was the greatest gift I could imagine…
          What a darling film—everyone adored it.
        4.If you describe someone as a darling, you are fond of them and think that
          they are nice. [INFORMAL]
          He's such a darling. 5. The **darling** of a group of people is someone who
          is especially liked by that group.
          Rajneesh was the darling of a prosperous family.

One can see that there is a cultural difference between the usages of honey in American English and darling in British English, furthermore, there are differences in the use of darling among British people as well.

In addition to the meanings enumerated in the COBUILD dictionary (Sinclair 2001), it can be seen that the use of *Darling* may convey ironic or sarcastic content. I have browsed some forums on the Internet regarding what English speaking language users think about the use of *Honey* and *Darling* and whether there is any difference between them. Let us see some opinions taken from the discussion

---

[7] I am indebted to my colleague József Andor who drew my attention to the cultural differences of English vocative forms in a personal conversation.

about *honey* and *darling* on the website Wordreference.com Language Forums http://forum.wordreference.com/threads/honey-vs-darling.1245116/ (date of access July 16, 2015).

(10) "Darling" can also be used as an adjective (including somewhat ironically): Well, that was my **darling** sister on the phone. Her car has developed engine trouble, she is stranded, and so she expects me to drive 100 miles to come and pick her up.(GreenWhiteBlue, Jan 20, 2009)

(11) Funny how this varies so much culturally. In Barbados, honey, sweetie, sweetheart and darling are both used quite normally and naturally as terms of endearment. They can be used romantically, casually between friends, family members, between acquaintances, or even strangers (people in the service industry and customers for example). Personal preference will determine which word is chosen and context will determine how the word is interpreted.

I would not say that they are entirely interchangeable, but close enough. Older people tend to use "darling", whereas "honey" is more common among younger people. Neither is typically used in a tongue-in-cheek manner here ewie and Loob.

Cicciosa, I advise that you find out how these words are used in the **particular** cultural context in which you intend to use them. As you can see from the posts so far, there is no "correct" answer to the question and usage varies widely. (Nymeria, Jan 20, 2009)

(12) I consider "darling" a much stronger term of endearment than "honey". Although some people in some circumstances use "darling" indiscriminately —as, perhaps, a waitress to a customer—I would only use the word with a lover or a close relative. "Honey" is for anyone. (cyberpedant, Jan 20, 2009)

(13) Honey and darling could be used between lovers (I suppose you meant people who have sexual intercourse together), but they are quite common in film and TV for quite innocent male-female relationships. As to UK English darling is/was common and honey reached us via the outpourings of films and TV j…from over the pond, I believe…(George French, Jan 20, 2009)

On the basis of the COBUILD dictionary (Sinclair 2001), my observations as well as forum comments and examples of the commenters, it can be said that in some English speaking cultures or subcultures there is a difference between the language users in the use of *honey* and *darling*. In an English speaking culture where there is a difference between the use of vocative forms Honey and Darling it conveys important information which form a speaker selects. The cultural knowledge concerning the use of these terms constrains the use of Honey and Darling in a particular context and the use of these terms indicates to the hearers what perspectives and intentions the speakers have. Let us turn back to (5) which I repeat now as (14) for convenience.

(14) Can Billie have an ice-cream, Darling?

Let us imagine that the wife does not use the vocative form *Darling* in normal circumstances, she only uses this form when she is not satisfied with her husband's behavior to indicate him her dissatisfaction. When she realizes that her husband is eating ice-cream, but he does not give any to Billie she becomes furious. She tries to express her anger and dissatisfaction by using *Darling*. Hearing *Darling*, the husband may infer that his wife formulates her utterance from her own perspective as well as her intention to order (and not request) the husband to give Billie ice-cream. Although (14) also has the conventional linguistic form of an indirect request, the particular context contains a piece of information about the wife's anger and dissatisfaction which overrides the interpretation as indirect request predicted by the general pragmatic knowledge, instead, it supports the interpretation as an order. The wife's intentions and her cultural knowledge about their family life motivate her to select the form *Darling*. The form *Darling* creates a particular context for the husband and he recalls his cultural knowledge about their family life and his wife's habits of using *darling* when she is dissatisfied. Taking into account these various kinds of information, the husband may infer her wife's communicative and informative intentions as well as her intention to order to give Billie ice-cream. Let us add that a dynamic interaction can be presumed both in the use and interpretation of the cultural knowledge, general pragmatic knowledge, particular context and linguistic form of the utterance (cf. also Kecskes 2014; Németh T. and Bibok 2010).

1.4 Summary

Verbal communication may be successful if the communicative partners can take into account each other's perspective and formulate a shared perspective. Communicators attempt to provide clear clues about which perspectives they should assume and how they should move from one perspective to the next (MacWhinney 2005, 198). The perspectives of communicative partners include their informative and communicative intentions (Németh T. 2014, 2015). The content of communicators' informative intention may contain various other intentions such as to ask, to request, and to show one's appreciation.

As we saw in the examples in the previous section, the speakers' perspectival intentions in communication can be expressed by various linguistic and pragmatic devices within perspectives. The linguistic indicators of intentional viewpoints may be thought of as a special kind of deixis, it can be called intention deixis, similar to person, time or social ones. I have identified some linguistic (e.g. forms of vocative selection, question form for indirect request) and contextual clues (asking gaze, pieces of information from the general pragmatic, cultural knowledge as well as particular contextual information) representing the speakers' intentions. The complex analysis suggested in this chapter can be extended to the informative and manipulative forms of language use.

In informative language use, the speaker has only an informative intention without communicative intention. In order to convey information, the speaker also wishes to attract the attention of a listener, but (s)he cannot provide explicit linguistic clues, (s)he cannot use a vocative form or perform an information requesting question or an indirect request (cf. Németh T. 2008, 2014, 2015). Instead, (s)he can communicate with a partner and should speak loud enough for another listener to hear her/his utterance. In informative language use the speaker wishes to develop a perspective in the person to be informed which does not contain any information about her/his informative intention. In informative language use the speaker does not want the person to be informed to take her/his perspective entirely, thus the speaker's and hearer's perspectives should not coincide completely.

In verbal manipulation the speaker has a manipulative intention in addition to the informative intention in manipulative information transmission or the informative and communicative intentions in verbal communication. Manipulative information transmission and manipulative communication may be successful if the manipulator's manipulative intention is not recognized by the partner. Therefore the manipulators attempt to avoid providing clear clues on the basis of which manipulative intention can be revealed. In successful manipulative language use, the intentional viewpoints of the communicator and her/his partner should not share the manipulative intention.

However, in order to successfully manipulate, a manipulator may apply several manipulative strategies. These strategies can be performed with the help of various linguistic (false presupposition, inclusive *we*, hyperbole, factive verbs such as *know* and *believe*) and pragmatic tools (e.g. false conversational implicatures). If the partners consciously interpret the utterances they can realize these elements, and they may serve as clues to reveal manipulative intention (Árvay 2004).

References

Abuczki A (2011) A multimodális kommunikáció szekvenciális elemzése [sequential analysis of multimodal communication]. In: Enikő NT (ed) Ember-gép kapcsolat. A multimodális ember-gép kommunikáció modellezésének alapjai [The relationship between human and machine. The bases of the modelling of multimodal communication between human and machine]. Tinta Könyvkiadó, Budapest, pp 119–144

Abuczki A, Bódog A, Németh T. E (2011) A multimodális pragmatikai annotáció elméleti alapjai az ember-gép kommunikáció modellálásában [the theoretical basis of multimodal pragmatic annotation in the modeling of human-machine communication]. In: Németh T. E (ed) Ember-gép kapcsolat. Tinta Könyvkiadó, Budapest, pp 179–201

Ariel M (2008) Pragmatics and grammar. Cambridge University Press, Cambridge

Árvay A (2004) Pragmatic aspects of persuasion and manipulation in written advertisement. Acta Linguist Hung 51:231–263

Bezuidenhout A (2013) Perspective taking in conversation: a defense of speaker non-egocentricity. J Pragmat 48:4–16

Buda B (1988) A közvetlen emberi kommunikáció szabályszerűségei [Rules of direct human communication]. Tömegkommunikációs Kutatóközpont, Budapest

Bunt H, Black W (2000) Abduction, belief and context in dialogue. Studies in computational pragmatics. Benjamins, Amsterdam

Ford CE, Thompson S (1996) Interactional units in conversation. Syntactic, intonational and pragmatic resources for the management of turns. In: Ochs E, Schegloff EA, Thompson S (eds) Interaction and grammar. Cambridge University Press, Cambridge, pp 134–184

Goodwin C (1981) Conversational organization. Interaction between speakers and hearers. Academic Press, New York

Grice HP (1957) Meaning. Philos Rev 66:377–388

Grice HP (1975) Logic and conversation. In: Cole P, Morgan JL (eds) Syntax and semantics volume 3. Speech acts. Academic Press, New York, pp 134–184

Haugh M (2010) Co-constructing what is said in interaction. In: Németh T. E, Bibok K (eds) The role of data at the semantics- pragmatics interface. de Gruyter, Berlin, New York, pp 349–380

Horn LR (2004) Implicature. In: Horn LR, Ward G (eds) The handbook of pragmatics. Blackwell, Oxford, pp 3–28

Hunyadi L (2011) Multimodal human-computer interaction technologies. Theoretical modeling and application in speech processing. Argumentum 7:240–260

Hunyadi L (2013) Possible communicative cues to syntactic incompleteness in spoken dialogues. Argumentum 9:186–199

Ivaskó L, Németh T. E (2002) Types and reasons of communicative failures: a relevance theoretical approach. Modern Filológiai Közlemények 4:31–43

Jakobson R (1960) Closing statements: linguistics and poetics. In: Sebeok TA (ed) Style in language. MIT Press, Cambridge Massachusetts, pp 350–377

Jurafsky D (2004) Pragmatics and computational linguistics. In: Horn LR, Ward G (eds) The handbook of pragmatics. Blackwell, Oxford, pp 578–604

Kecskes I (2014) Intercultural pragmatics. Oxford University Press, Oxford

Levinson SC (1983) Pragmatics. Cambridge University Press, Cambridge

Lewis DK (1969) Convention. The MIT Press, Cambridge, MA

MacWhinney B (2005) The emergence of grammar from perspective. In: Pecher D, Zwaan RA (eds) The grounding of cognition: the role of perception and action in memory. Cambridge University Press, Cambridge, pp 198–223

Németh T. E (1990) Az emberi kommunikáció kutatásának néhány alapkérdése [some basic questions of the research into human communication]. Néprajz és Nyelvtudomány 33:43–56

Németh T. E (1996) A szóbeli diskurzusok megnyilatkozáspéldányokra tagolása [Segmentation of spoken discourses into utterance tokens]. Akadémiai Kiadó (Nyelvtudományi Értekezések 142.), Budapest

Németh T. E (2005) Az osztenzív-következtetéses kommunikációtól a verbális kommunikációig [from ostensive-inferential communication to verbal communication]. In: Ivaskó L (ed) Érthető kommunikáció [Understandable communication]. SZTE, Médiatudományi Tanszék, Szeged, pp 77–87

Németh T. E (2008) Verbal information transmission without communicative intention. Intercult Pragmat 5:153–176

Németh T. E (2011) A humán kommunikáció modelljei és az ember-gép kommunikáció [models of human communication and human-machine communication]. In: Németh T. E (ed) Ember-gép kapcsolat. A multimodális ember-gép kommunikáció modellezésének alapjai [The relationship between human and machine. The bases of the modelling of multimodal communication between human and machine]. Tinta Könyvkiadó, Budapest, pp 43–61

Németh T. E (2014) Intentions and perspectives in the social forms of language use. Argumentum 10:472–485

Németh T. E (2015) The role of perspectives in various forms of language use. Semiotica 203:53–78

Németh T. E, Bibok K (2010) Interaction between grammar and pragmatics: the case of implicit arguments, implicit predicates and co-composition in hungarian. J Pragmat 42:501–524

Sanders J, Spooren W (1997) Perspective, subjectivity, and modality from a cognitive linguistic point of view. In: Liebert WA, Redeker G, Waugh L (eds) Discourse and perspective in cognitive linguistics. John Benjamins Publishing Company, Amsterdam/Philadelphia, pp 85–112

Shannon C, Weaver W (1949) The mathematical theory of communication. University of Illinois Press, Urbana, IL

Sinclair J (2001) Collins COBUILD English dictionary for advanced learners, 3rd edn. Harper-Collins Publishers, Glasgow

Sperber D, Wilson D (1986/1995) Relevance. Cognition and communication, 2nd edn. Oxford, Blackwell

Toda M (1967) About the notions of communication and structure. In: Thayer L (ed) Communication. Concepts and perspectives. Spartan Books, Washington, London, pp 25–46

Tomasello M (1999) The cultural origins of human cognition. Harvard University Press, Cambridge, MA

Part II
Methods of Observation

Chapter 2
The Teacher's Body Communicates. Detection of Paraverbal Behaviour Patterns

Marta Castañer and Oleguer Camerino

Abstract The purpose of the present chapter is to describe how a systemic approach can be combined with the empirical detection of behaviour patterns by means of a systematic methodology and its utility of observing pedagogic communication. Of course, each teacher has his or her own paraverbal communicative style. However, the objective of this chapter is not to compare styles but, rather, to reveal the trends in this dimension of communication among teachers working in a similar naturalistic context. The observation of a natural context requires the use of the observational instrument, as well as the detection of temporal patterns in the transcribed actions. Therefore, despite the concrete and unique nature of each body it is possible to identify certain kinesic and proxemic functions and morphologies that are sufficiently generalised and which are of great interest with respect to teaching.

2.1 Introduction

The different languages used by human beings generate a peculiar system of signs that have their own specific semiotics. This gives these signs their singular nature and offers humans the possibility of a wide range of expression. In this regard, poets are creators of metaphors through which they represent gestures of any kind and the reality that surrounds us. An example of this was beautifully expressed by Virginia Woolf: *My spine is soft like wax near the flame of the candle*. If we treat our gesturality as a form of writing, then the body can be said to reveal itself. As a result, there is much to be uncovered by researchers. Indeed, we experience our cultures not only through discourse, signs and meaning, but also through the movements of

M. Castañer (✉) · O. Camerino
National Institute of Physical Education of Catalonia (INEFC), University of Lleida, Lleida, Spain

Lleida Institute for Biomedical Research Dr. Pifarré Foundation (IRBLLEIDA),
University of Lleida, Lleida, Spain
e-mail: mcastaner@inefc.es

O. Camerino
e-mail: ocamerino@inefc.es

© Springer Nature Switzerland AG 2020
L. Hunyadi and I. Szekrényes (eds.), *The Temporal Structure of Multimodal Communication*, Intelligent Systems Reference Library 164,
https://doi.org/10.1007/978-3-030-22895-8_2

our bodies. "Ways of behaving, of moving, of gesturing, of interacting with objects, environments, technologies, are all cultural" (Wise 2000: 303).

Bodies are self-sustaining systems (Jordan and Ghin 2006; Streeck and Jordan 2009) and "are naturally 'semiotic' in that they are natural representations of their embodied contexts. In a sense, they 'signify' the multiescale contexts they embody." (Streeck and Jordan 2009: 451). We believe that our approach and the findings of this chapter contribute to the work currently being carried out in body communication as a self-sustaining system. As pointed out many years ago by Goffman (1959), the stages on which the body 'moves' are always determined by coordinates of space and time, which are responsible for the contextualisation and evolution of our species. From a systemic point of view the body can be regarded as being inscribed upon continuous stages of space and time, on which multiple learning interactions take place in a flow-like manner.

Within these frameworks or stages of action we can distinguish three levels of interaction: With oneself (the inner world), with objects (the inanimate world) and with others (the animated/social world). Hence we are concerned with the capacity to act rather than to re-act, in other words, the ability to interact. This reality attributes to the body a singular nature amidst the multiplicity of 'images' that make up the universe in which we live, and each one of these three levels of interaction shapes a different concept of the body, namely: an identified body (with itself), an extended body (by means of objects and technology) and an objectified body (in relation to others).

Each of these forms of bodily existence is directly related to various dimensions set out in systemic approaches such as the organismic system theory of Ludwig von Bertalanffy (1969). These dimensions are, respectively, the *introjection*, the *extension* and the *projection*, and they characterise the intelligent human system that is capable of generating multiple and singular modes of *symbolisation* and *codification*, this being the origin of language and the different forms of human communication.

2.1.1 Introjection, Extension and Projection of the Teaching Discourse

In order to illustrate the *introjection dimension*, the phenomenology of the body from Merleau-Ponty (1962) to Michel Foucault (1982) helps us to avoid a restrictive view of introjection. Hence, this dimension can be contemplated in a wider sense, ranging from Merleau-Ponty (1962) concept of the lived body (corps vécu) to Foucault (1982) notion of the body as the product of cultural practices. The former alludes to the perceptual potential of the body and its capacity for action that enables it to open itself to the external world, whereas the latter alludes to the fact that the body is shaped by the various bodily constructions that to use the language of the computer age format it. The former notion is pre-conceptual and pre-cultural, and allows the body to be referred to in the first person, this being consistent with the use of reflexive verbs such as 'to recognise oneself'. By contrast, the latter is conceptual and cultural, and allows the body to be referred to in the third person, which, as we shall see, is an aspect that is directly related to the projection dimension.

In the kinesic communication of the body this aspect can be witnessed on a daily basis in the morphology of gestures, and each society, each professional group and, therefore, each individual teacher will produce a particular set of gestures (Pozzer-Ardenghi and Roth 2008). As such, the body acts like a crucible, a site in which cultural constructions are filtered and a communicative language (both kinesic and proxemic) is developed that both reflects these constructions and influences every process of teaching and learning.

To illustrate the *extension dimension* of the body, Merleau-Ponty (1962) described how the world of objects (which also includes technology) is incorporated into our bodies. He does this by means of two examples: That of a blind man's cane, through which his body can be extended and which, to a certain extent, becomes part of his body, and secondly, that of a woman's feathered hat, which also extends her body but without having to be manipulated as in the case of the walking cane. These are quasi-extensions of the body that show how the material nature of technology and the tactile aspect of our sensoriality underlie the human body's great potential for extension in the social and three-dimensional world (Goldin-Meadow 2003).

In the body's kinesic communication, this extension is made possible through the adaptive gestures that the teacher makes when coming into contact with objects or, at times, the bodies of other people. However, given that they are produced unconsciously these adaptive gestures are usually a form of extra communication, whereas the real power of extending our communication this ways comes, paradoxically, from gestures that do not require any kind of object, i.e. deictic gestures that have their origins in the primordial gesture of those hominids who first used their hand for indicative purposes.

In order to illustrate the idea of the *projection dimension*, Heidegger (1982) uses classical phenomenology to show how the use of objects allows humans to project themselves into their work practices. This, therefore, provides an interesting way of illustrating this idea in crescendo, which goes from the introjection to the projection dimension as the projection aspect entails interpersonal relationships creation.

As regards the objectified body, Sartre, in *Being and Nothingness*, considered the power of the gaze that gives rise to the conflict between seeing and being seen by an eye that objectifies interpersonal relationships. However, more than just a conflict we regard this as a positive tension, since the negentropy in human relations is achieved by establishing (and simultaneously regulating) the tensions between opposing aspects. Becoming an object in the gaze of the other is one such aspect, as discussed by Marcel in *Being and Having*, where he highlights the mutually participative nature of this objectification in human relationships. Here, it should not be forgotten that teacher and pupil are also two bodies that, in every context of face-to-face teaching, repeatedly observe each other.

2.1.2   The Non-linearity of Human Movement

The body is what one sees, yet human movement vanishes in our everyday perception. Increasingly we need to understand how the geometry of our bodies is radiated and expressed in relation to others in any act of communication, including academic teaching practices. Speech can be viewed as the style of a given individual, in line with the idea of Italo Calvino (1974) when he said that signs create a language but not the language we know. Each language has its specific form of expression that allows an exhaustive taxonomy to be established, but above and beyond any taxonomy, languages coexist and become interwoven in a highly complex game. The text of the body has never been linear in the sequential sense. Its gestural kinesics and proxemics, or the use of space, all constitute constraints emerging in the majority of contexts.

Spoken language is usually imbued with a 'tone' that is embedded in a form of expressivity that transcends the verbal sphere, in line with what neurologists such as Oliver Sacks have discussed in their writings. This expressivity is spontaneous and, as such, cannot be easily faked in the way that words can be. As *Homo loquens*, human beings are able to specify what we could call the hidden meaning of words. "One can lie with the mouth", Nietzsche (1954) writes, "but with the accompanying grimace one nevertheless tells the truth". When the language used is derived from corporality and is also the object of study, one is faced with the paradox of understanding corporality as the *language of silence*.

Discourse is transformed into a series of movements within language in such a way as to give it meaning, and hence the body is revealed as a piece of writing. In this context, one must consider the semiotics of the body (Streeck and Jordan 2009; Lemke 2000), that of the res *extensa*, often translated as 'corporeal substance' by Descartes and whose textuality paves the way for the interpretations made by the reader who observes human movement. Following Foucault (1982) this provides a new and living *episteme* for semiotics that, for human movement, is enormously rich and communicative and, as such, revealing.

2.1.3   Paraverbal Communication and Body Language

Some literature reviews are organised around conceptual and methodological approaches used in the study and applications of non-verbal behaviour (Wolfgang 1997). At any rate we think that it is important to clarify an aspect related to non-verbal and paraverbal concepts. In our view the use of the negative prefix implies that the terms 'verbal' and 'non-verbal' should be understood as being mutually exclusive, when in fact they refer to two forms of communication that go hand in hand with one another. Indeed, we experience our culture not only through discourse, signs and meaning, but also through the movements of our bodies. Paraverbal teaching style refers to the ways in which a teacher conveys his or her educational discourse, and this is why it is sometimes associated with the idea of expressive movement (Gallaher

1992). De Vries et al. (2009) also define communicative style as the characteristic way a person sends verbal, paraverbal, and non-verbal signals in social interactions.

According to Gadamer (1980) good understanding lays not so much in listening to things said by others, but rather listening to ourselves in relation to others, and the same applies to the processes of seeing ourselves and being seen in relation to our body language. Thus, paraverbal communication is subject to certain social norms regarding gestural configurations (Roth 2001), both kinesic (Birdwhistell 1970; Kendon 1993) and proxemic (Hall 1968), which cannot exist outside the ethno-aesthetics of a given historical moment. In this context, kinesics is the study of patterns in gesture and posture that are used with or without communicative meaning, while proxemics is the study of how we use space in academic teaching practices.

These dimensions can appear simultaneously or concurrently, functioning in an integrated and systemic way. If communication is to be effective, it is necessary to ensure that all the paraverbal dimensions are congruent, i.e. that they seek to transmit the same message, strengthening, confirming and heightening it in accordance with the educational circumstances (Jones and LeBaron 2002). The present study focuses on the paraverbal dimensions of proxemics and kinesics, and below be provide a more detailed conceptual description of both of these.

2.1.4 From Kine to Gesture

At this point it seems relevant to clarify a conceptual aspect that continues to be overlooked in the area of kinesic language based on human motor behaviour. Firstly, it is necessary to distinguish between kine, posture, gesture and attitude associated with the body (Castañer et al. 2010, 2016). Kine is the basic unit of movement, comparable to the phoneme of verbal language; body posture denotes the static nature of the body relative the position of its various osteoarticular and muscular parts, body gesture refers to the dynamic nature of the body, without forgetting that each gesture is comprised of multiple micro-postures; and body attitude is the meaning that each social group gives to the emotional and expressive ways of using postures and gestures.

The diverse, and at the same time, bilateral structure of our corporeity allows us to generate bodily postures (dynamism), gestures (dynamism) and attitudes (meaning) (Castañer et al. 2012) in a simultaneous way and also "gestures are often subsequently replaced by an increasing reliance upon the verbal mode of communications" (Roth and Lawless 2002: 285). Despite the concrete and singular nature of each body it is possible to identify certain kinetic and proxemic functions and morphologies that are sufficiently generalised, and which are of great relevance to the process of teaching in the academic context, this being the aim of our research line. On the basis of this initial clarification, gesture can be regarded as the basic unit of meaning for constructing the paraverbal kinesic observational system. Consider the chart shown in Fig. 2.1.

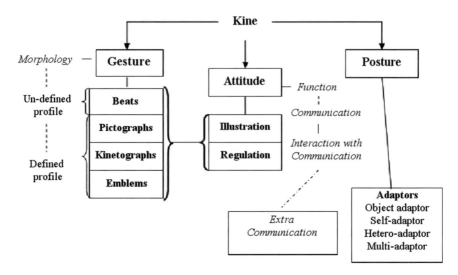

Fig. 2.1 Relationship between the morphology and function of kinesic gestures

As regards the *morphology* of the categories (see Fig. 2.1) we will define a continuum that encompasses: gestures that offer a highly-defined profile and which are clearly observable by the recipient and gestures with a less well-defined and weaker profile.

A clear example of those kine that offer a well-defined gesture profile, are the emblem gestures (Ekman 1985). With regard to their *functionality* we establish a continuum that encompasses: Gestures with a purely communicative purpose, gestures whose purpose is communication with interaction and extra-communicative gestures, i.e. those without any explicit interactive or communicative purpose.

It should be mentioned that the structure presented is clear enough for carrying out empirical research of the kind proposed herein, not least because it is based on a long scientific tradition. For example, Wiener et al. (1972) made a distinction between gestures that have a communicative function and those that do not. This distinction has certain relevance in our present study. In their paper Wiener et al. argued that communicative gestures comprise those of a *pantomimic* kind (highly stylised and defined improvised movements that represent an object or an event), and the majority of gestures that *accompany words* in their semantic itinerary in order to reinforce the relationship between sender and receiver. Furthermore, they note that kinesic movements of the adaptor kind do not have a communicative function. As for functional description there are other proposals such as that of Cosnier and Brossard (1984), who characterised six types of movement: *Quasi-linguistic* (equivalent to the emblems of Ekman and Friesen (1969), *expressive, regulatory* (organisation of social interaction), co-verbal (everything that can be considered as an illustrative gesture), *meta*-communicative, and *extra-communicative* (changes in posture, self-manipulation, object manipulation). Efron (1972) identified two kinds of gestures that were clearly linked to the expository process: Beats, which accompany the melodic aspects and rhythm of language, and *ideographs*, whose function is to describe everything that appears in thought. Subsequent research, mainly the result of earlier work by Ekman and Friesen (1969) and Kendon (1969), was founded on the triple functionality of kinesic gestures that is given by their *origin* (innate, natural or cultural), their *coding* (arbitrary, iconic or intrinsic) and their *use*. These authors also distinguished, among other categories, between illustrative and regulatory gestures. Thus, the ongoing research of investigators such as Kendon (1993) in this area has explored further the kinesic repertory and offers a description of different types of gestures (deictic, beats, iconic and metaphoric), because body gestures are always an integral component of language (McNeill 2005).

Although the adequate use of any source of illustration can foster learning, it is worth noting the conclusion reached by a recent study about the effect of illustrations in arithmetic problem-solving: "The results show that illustrations can have a detrimental effect on performance in arithmetic word problems, produced by irrelevant, redundant or interacting sources of information" (Berends and van Lieshout 2009: 345). Paraverbal behaviour is largely unconscious and needs to be made conscious in order to optimise it. The proximity between teachers and students can be perceived by means of gaze, gestures and spatial location, all of which have an affective component and which can influence the intrinsic motivation felt toward the material and the educational setting (Rodriguez et al. 1996). As such it is feasible to achieve more effective paraverbal communication in accordance with the objectives being sought.

An intrinsic part of all teaching activity is a constant communicational flow, in which the spontaneous nature of communication is considered to be a habitual feature; Buck and VanLear (2002) even went as far as to define this as non-intentional communication. The observation of students' reactions may thus be useful for optimising this communication (Moore 1996). As such, there is good cause why communication is regarded as an indicator of the communicator's emotional, as well as symbolic experiences (Le Poire and Yoshimura 1999). Symbolic communication is intentional communication that uses learned, socially-shared signal systems of propositional information transmitted via symbols. Furthermore, it should also be added that "gestures support the development of verbal modes by decreasing the mental effort required for producing communication" (Roth 2004: 2). Thus, observational methodology is used due to the habitual nature of teachers' behaviour and the fact that the context is a naturalistic one. The flexibility and rigour of this methodology makes it fully consistent with the characteristics of the study and it has become a standard approach in observational research, especially in the area of kinesic and paraverbal communication (Izquierdo and Anguera 2001; Castañer et al. 2013, 2016).

2.2   Methods

2.2.1   Pattern Analysis and the Systemic Approach

In order to improve the scenarios to be managed in academic teaching practices, it is important to identify the essential aspects of communication such as gestures, voice quality and the use of teaching time and space which are associated with the teaching discourse. In this regard, it is clear that one of the keys in optimising academic teaching practices lies in paying close attention to how communicative and teaching styles are reworked over time. Through the detection of temporal patterns (T-patterns) we can observe and analyse all these pedagogical constraints, and this rigorous analytic procedure provides a holistic point of view that is consistent with the systemic approach taken so far. T-patterns can be detected and analysed with the *Theme* v.5 software (Magnusson 1996, 2000, 2005; Pattern Vision 2001). *Theme* not only detects temporal patterns but also indicates the relevance and configuration of recorded events. The approach is based on a sequential and real-time pattern type (T-patterns), which, in conjunction with detection algorithms, can describe and detect behavioural structure in terms of repeated patterns. It has been shown that such patterns, while common in behaviour, are typically invisible to observers, even when assisted by standard statistical and behaviour analysis methods. The T-pattern algorithm is implemented in the specialised software package, *Theme* (see www.patternvision.com and www.noldus.com). *Theme* also displays event frequency charts based on the occurrences of recorded events and the frequency of each category independently of the other categories. The detection of T-patterns has proven to be extraordinarily productive and fruitful for the study of the multiple facets or types of body movement (Camerino et al. 2012; Sakaguchi et al. 2005), as well as for non-verbal communication (Blanchet et al. 2005; Haynal-Reymond et al. 2005; Castañer et al. 2013, 2016), sport (Borrie 2001; Borrie et al. 2002; Bloomfield et al. 2005; Jonsson et al. 2006) and motor skills (Castañer et al. 2018, 2009; Casarrubea et al. 2018).

Our main line of research is based on observational methodology with the aim of identifying the kinesic and proxemic patterns used in discourse that are not strictly verbal. Our intention is not to explore in depth the hidden dimensions of academic discourse, but rather to study what is directly observable from an objective point of view.

2.2.2   Participants

We recorded classroom-based lessons on various subjects and taught by three experienced teachers offering pre-university courses. A total of twelve sessions (four lessons taught by each teacher) were analysed. Although, in this study, we obtained various data about the communicative style of each teacher, we were, in fact, only focused on on identifying the overall communicative style of the teachers.

2.2.3 Instruments

The observation tool used was SOCOP, which allows the different levels of kinesic and proxemic response to be systematically observed. Kinesic responses were recorded by means of the Sub-system for the Observation of Kinesic Gestures (SOCIN; see Table 2.1), while proxemic gestures were recorded via the Sub-system for the Observation of Proxemics (SOPROX; see Table 2.2). Both sub-systems were successfully used in a previous study of observing the behaviour of teachers interacting with their students (Castañer et al. 2010, 2012, 2013, 2016).

We think that this tool offers greater applicability and flexibility than do other existing tools which, in our view, are hindered by a degree of analysis that is too complex; for example, the kinesic analyses of Birdwhistell (1970) in the field of non-verbal human communication, or the notation systems of Laban and Ullman (1988) provide a considerable amount of information but they are very difficult to use in many natural contexts where communicative teaching might be observed.

The SOCIN tool, for kinesic actions, according to the theoretical framework we have made above, is based on four variables (morphology, function, adaptor and situation). Similarly, the SOPROX tool, for proxemic actions, is based on five variables (group, topology, location, orientation and transition). Observational methodology requires a clear and exhaustive definition of each of the categories included in the observation system or field format. Each of the criterion, categories and codes that form part of the SOCIN (Table 2.1) and SOPROX (Table 2.2) tools are defined below.

2.2.4 Materials and Procedure

The recording tool used to codify SOCOP was the LINCE program (Gabin et al. 2012), which was constructed as a software package that automates the functions of the design of observational systems, video recording, the calculation of data quality and the presentation of results which can be exported in various formats, those of THEME, GSEQ, EXCEL and SAS. Sessions are digitised to make them available for frame-to-frame analysis and enable them to be coded in the LINCE program (Fig. 2.2). In all our sessions the behaviour of teachers is always observed continuously. The procedure was in line with APA ethics and was approved by the university departments involved. The project did not involve any experiments or manipulation of subjects. The results are based on data obtained from recordings of classroom sessions, but in line with the Belmont Report (National Commission for the Protection of Human Subjects of Biomedical and Behavioral Research, 1979) in order to assure that the subjects' rights have been respected. As such, the photo images shown in this chapter were created for the express purpose of illustration, not representing original persons.

2.3   Results and Discussion

As regards the criteria defined by the SOCOP observation tool the results allow us to highlight a series of trends in both kinesic and proxemic communication, and also combinations of the two. The *Theme* program derived T-patterns (temporal patterns) that reveal the trends in kinesic and proxemic paraverbal communication from an ideographic perspective. As an example, let us consider a T-pattern that is of interest for to the generation of paraverbal communicative responses. Figure 2.3 shows the most complex T-pattern discovered from all the observational data files we have so far.

The *Function criterion* reveals that most teachers use the regulatory function 30% of the time, the remaining 70% corresponding to the illustrative function; in other words, actions that do not require an immediate response such as explaining or providing information account for the largest proportion of time less for regulatory actions, the latter expecting an interaction or response such as asking questions, giving orders and offering help.

Table 2.1 SOCIN: System of Observation for Kinesic Communication. (Castañer et al. 2013)

Dimension	Analytical categorization	Code	Description
Function It refers to the intention of the spoken discourse that the gesture accompanies	Regulatory	RE	Action by the teacher whose objective is to obtain an immediate response from recipients. It comprises imperative, interrogative, and instructive phrases with the seek of exemplifying, giving orders or formulating questions and answers
	Illustrative	IL	Action that does not aim to obtain an immediate response from the recipients (although possibly at some future point). It comprises narrative, descriptive and expository phrases with the aim of getting receivers to listen
Morphology It refers to the iconic and biomechanical form of gestures	Emblem	EMB	Gesture with its own pre-established iconic meaning
	Deictic	DEI	Gesture that indicates or points at people, places or objects
	Pictographic	PIC	Gesture that draws figures or forms in space

(continued)

Table 2.1 (continued)

Dimension	Analytical categorization	Code	Description
	Kinetographic	KIN	Gesture that draws actions or movements in space
	Beats	BEA	Iconically undefined gesture used exclusively by the sender and which usually only accompanies the logic of spoken discourse
Situational It refers to a wide range of bodily actions which usually coincide with parts of the teaching process that cover a certain period of time	Demonstrate	DE	When the teacher performs in gestures that which he or she wishes the students to do
	Help	HE	When the teacher performs actions with the intention of supporting or improving the contributions of students
	Participate	PA	When the teacher participates alongside students
	Observe	OB	Period of time during which the teacher shows an interest in what is happening in the classroom with the students
	Provide material	PM	When the teacher handles, distributes or uses teaching material in accordance with the educational setting
	Show of affect	AF	When the teacher uses an emotionally-charged gesture with respect to the students
Adaptation It refers to gestures without communicative intentionality in which the teacher makes contact with different parts of their body, or with objects or other people	Object adaptor	OBJ	When the teacher maintains contact with objects but without any communicative purpose
	Self-adaptor	SA	When the teacher maintains contact with other parts of their body but without any communicative purpose
	Hetero-adaptor	HA	When the teacher maintains bodily contact with other people but without any communicative purpose
	Multi-adaptor	MUL	When several of these adaptor gestures are combined

Table 2.2 SOPROX: system of observation for proxemic communication. (Castañer et al. 2013)

Dimension	Analytical categorization	Code	Description
Group It refers to the number of students to whom the teacher speaks	Macro-group	MAC	When the teacher speaks to the whole class/group
	Micro-group	MIC	When the teacher speaks to a specific sub-group of students
	Dyad:	DYA	When the teacher speaks to a single student
Topology It refers to the spatial location of the teacher in the classroom	Peripherial	P	The teacher is located at one end or side of the classroom
	Central	C	The teacher is situated in the central area of the classroom
Interaction It refers to the bodily attitude which indicates the teacher's degree of involvement with the students	At a distance	DIS	Bodily attitude that reveals the teacher to be absent from what is happening in the classroom, or which indicates a separation, whether physical or in terms of gaze or attitude, with respect to the students
	Integrated	INT	Bodily attitude that reveals the teacher to be highly involved in what is happening in the classroom, and in a relation of complicity with the students
	Tactile contact	TC	When the teacher makes bodily contact with a student
Orientation It refers to the spatial location of the teacher with respect to the students	Facing:	FAC	The teacher is located facing the students, in line with their field of view
	Behind:	BEH	The teacher is located behind the students, outside their field of view
	Among:	AMO	The teacher is located inside the space occupied by the students
	To the right	RIG	The teacher is located in an area to the right of the classroom and of the students, with respect to what is considered to be the facing orientation of the teaching space

(continued)

Table 2.2 (continued)

Dimension	Analytical categorization	Code	Description
	To the left	LEF	The teacher is located in an area to the left of the classroom and of the students, with respect to what is considered to be the facing orientation of the teaching space
Transitions It refers to the body posture adopted by the teacher in space	Fixed bipedal posture	FB	The teacher remains standing without moving
	Fixed seated posture	FS	The teacher remains in a seated position
	Locomotion	LOC	The teacher moves around the classroom
	Support	SU	The teacher maintains a support posture by leaning against or on a structure, material or person

Fig. 2.2 Screen capture of LINCE. (Gabin et al. 2012)

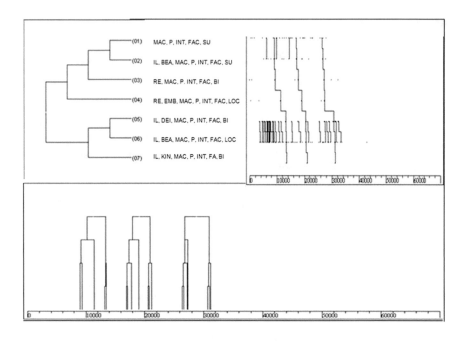

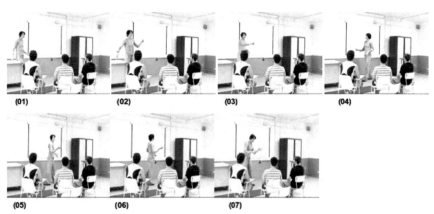

Fig. 2.3 (Continued)

◄ **Fig. 2.3** This given T-pattern is one of the most complex of those detected. It consists of six levels and a sequence of seven events, each one of which is composed of a complex combination of codes (combinations formed by between five and seven codes), occurring on three occasions during the observation period with the same sequence of events and significantly similar time intervals between each event occurrences. The interpretation that can be derived from the seven steps of this T-pattern sequence can be described step by step as follows: (01) The teacher interacts with the whole group/class, she is located at the periphery of the classroom (P) and facing the group (FAC) with her body position supported by a table or chair (SU); she displays an integrative (IN) attitude, participating in the activity being carried out by the students. (02) Next, the teacher offers an explanation using the illustrative function (IL), making gestures in the form of beats (BEA) and maintaining her orientation with respect to the group, as well as the spatial location described in (01). (03) She continues to maintain this orientation, her position shifts from fixed bipedal (BI) and she ceases to illustrate in order to make a self-adaptor gesture. (04) She goes on facing the whole class and at the periphery, but ceases to remain still in a bipedal position and begins to move around (LOC), without speaking, although she does make use of an emblem gesture (EMB) with a regulatory function (RE). (05) She returns to a combination (02), but this time in a fixed bipedal position and uses deictic gesture (DEI), and in (06) she maintains this but then shifts to locomotion (LOC). (07) She returns to a combination (05), but this time, instead of deictic gestures (DEI) she uses a kinetographic gesture (KIN) that displays a given action of movement in space. The duality formed by (05) and (06) was found to be very frequent. In fact, these two combinations have identical codes, although referring to alternating bipedal positions (FB) of the teacher with periods of movement (LOC)

Concerning the combination of the criteria *Morphology* and *Function* of gestures it can be seen that emblems, deictic forms, pictographs, kinetographs and beats are used without distinction in order to convey each function, whether it be regulatory or illustrative; however, gestures that are less well-defined in terms of morphology, such as beats, are more likely to accompany the illustrative function, whereas most emblems and deictic forms, both of which are gestures with a well-defined morphology, tend to accompany more the regulatory function. In our view the *Adaptation* criterion is of less interest to us as it refers to extra communicative aspects associated with unconscious contact gestures made by the sender shown by their high frequency.

As for the *Transitions* criterion, fixed bipedal postures are usually alternated with periods of locomotion as the teacher moves from one area of the classroom to another. Occasionally one can observe support postures, generally in conjunction with tables or chairs, but when posture is static in the seated position this tends to be maintained for some time.

Concerning the relationship between the *Function* and *Transitions* criteria the results suggest a common association between the regulatory function and static bipedal postures, whereas the illustrative function is combined with locomotion or movement around the classroom. It appears that when giving an illustration, which does not require a gesture of interaction, the teacher feels freer to move around. In contrast, the regulatory function, which does call for gestures indicating interaction seems to require greater concentration on the part of a teachers and leads them to fix their posture and thus focus their vision on a single point while asking questions, making comments or giving orders.

With the *Orientation* criterion the predominant position tends to be facing the group. Teachers rarely take up a position behind the group. The *Group* criterion tells us that interaction mostly occurs with the whole group, followed by that with micro-groups and, occasionally, with dyads.

2.4 Conclusion

In this chapter, our purpose was to describe how a systemic approach can be merged with the empirical detection of behaviour patterns by means a systematic methodology and its utility of observing pedagogic communication. As regards the verbal and paraverbal communication of teachers the introjection, extension and projection dimensions of the systemic approach presented are reflected in the communicative style of each individual teacher. More specifically, it can be seen in how he or she uses the functions of illustration and regulation, as well as in the meanings of the kinesic and proxemic repertories that are employed.

Various interlinked body gestures may convey the idea of a short sentence, but they do not have the scope achieved by, for example, the language developed specifically for deaf people. In this regard, mime, as the art of body language, does seek to produce a sentence, whereas interpersonal or pedagogic communication does not always do so. However, this should not be taken to mean that there is no grammaticality in body language; rather, the highly malleable nature of body language means that it is circumscribed in a diversity of human communication, including pedagogic contexts, with which it acquires different levels of meaning as a self-sustaining system (Jordan and Ghin 2006; Streeck and Jordan 2009). Hence, the fact that the embodied contexts associated with human communication require further analysis underlines the importance of paraverbal communication in teaching enhancing the predominant linear and figurative narrative, thus fostering a sort of kaleidoscopic patterns. We firmly believed that the temporal patterns we have detected can successfully optimise teacher discourse.

Acknowledgements We gratefully acknowledge the support of National Institute of Physical Education of Catalonia (INEFC) and the support of a Spanish goverment subproject Integration ways between qualitative and quantitative data, multiple case development, and synthesis review as main axis for an innovative future in physical activity and sports research (PGC2018-098742-BC31) (Ministerio de Economía y Competitividad, Programa Estatal de Generación de Conocimiento y Fortalecimiento Científico y Tecnológico del Sistema ICDCi), that is part of the coordinated project New approach of research in physical activity and sport from mixed methods perspective ($NARPAS_MM$) (SPGC201800X098742CV0); and the support of the Generalitat de Catalunya Research Group, Research group and innovation in designs (GRID). Technology and multimedia and digital application to observational designs (Grant No. 2017 SGR 1405).

References

Berends IE, van Lieshout E (2009) The effect of illustrations in arithmetic problem-solving: effects of increased cognitive load. Learn Instr 19(4):345–353

Bertalanffy Lv (1969) General system theory. George Braziller, New York

Birdwhistell R (1970) Kinesics and context. U.P.P, Philadelphia

Blanchet A, Batt M, Trognon A, Masse L (2005) Language and behavior patterns in a therapeutic interaction sequence. In: Anolli SD, Magnusson M, Riva G (eds) The hidden structure of social interaction. From Genomics to Culture Patterns. IOS Press, Amsterdam, pp 123–139

Bloomfield J, Jonsson GK, Polman R, Houlahan K, O'Donoghue PO (2005) Temporal pattern analysis and its applicability in soccer. In: Anolli SD, Magnusson M, Riva G (eds) The hidden structure of social interaction. From Genomics to Culture Patterns. IOS Press, Amsterdam, pp 237–251

Borrie A, Jonsson GK, Magnusson MS (2002) Temporal pattern analysis and its applicability in sport: an explanation and exemplar data. J Sport Sci 20:845–852

Buck R, VanLear CA (2002) Verbal and nonverbal communication: distinguishing symbolic, spontaneous, and pseudo-spontaneous nonverbal behavior. J Commun 52(3):522–541

Calvino I (1974) Invisible cities. Secker & Warburg, London

Camerino O, Castañer M, Anguera M (eds) (2012) Mixed methods research in the movement sciences: cases in sport, physical education and dance. Routledge, UK

Casarrubea M, Magnusson MS, Anguera MT, Jonsson GK, Castañer M, Santangelo A, Palacino M, Aiello S, Faulisi F, Raso G, Puigarnau S, Camerino O, Di Giovanni G, Crescimanno G (2018) T-pattern detection and analysis for the discovery of hidden features of behaviour. J Neurosci Methods 310:24–32

Castañer M, Torrents C, Anguera MT, Dinušova M, Jonsson G (2009) Identifying and analyzing motor skill responses in body movement and dance. Behav Res Methods 41(3):857–867

Castañer M, Camerino O, Anguera MT, Jonsson GK (2010) Observing the paraverbal communicative style of expert and novice PE teachers by means of SOCOP: a sequential análisis. Procedia Soc. Behav. Sci. 2(2):5162–5167

Castañer M, Andueza J, Sánchez-Algarra P, Anguera MT (2012) Extending the analysis of motor skills in relation to performance and laterality. In: Camerino O, ner MC, Anguera MT (eds) Mixed Methods Research in the Movement Sciences: Cases in Sport, Physical Education and Dance. Routledge, UK, pp 117–145

Castañer M, Camerino O, Anguera MT, Jonsson GK (2013) Kinesics and proxemics communication of expert and novice PE teachers. Quality Quant. 47(4):1813–1829. https://doi.org/10.1007/s11135-011-9628-5

Castañer M, Camerino O, Anguera MT, Jonsson GK (2016) Paraverbal communicative teaching t-patterns using SOCIN and SOPROX observational systems. In: Magnusson MS, Burgoon JK, Casarrubea MC (eds) Discovering Hidden Temporal Patterns in Behavior and Interaction. Springer, New York, pp 83–100. https://doi.org/10.1007/978-1-4939-3249-8

Castañer M, Andueza J, Hileno R, Puigarnau S, Prat Q, Camerino O (2018) Profiles of motor laterality in young athletes' performance of complex movements: merging the MOTORLAT and PATHoops tools. Front. Psychol. 9:916

Cosnier JY, Brossard A (1984) Communication non-verbale: co-texte ou contexte? Textes de base en psychologie: la communication non-verbale. Delachaux et Niestlé, París

De Vries RE, Bakker-Pieper A, Alting S, Van Gameren RK, Vlug M (2009) The content and dimensionality of communication styles. Commun. Res. 36(2):178–206

Efron G (1972) Gesture, race and culture. Mouton, The Hague

Ekman P (1985) Methods for measuring facial action. In: Scherer K, Ekman P (eds) Handbook methods in nonverbal behavior research. Cambridge University Press, Cambridge, pp 45–83

Ekman P, Friesen WC (1969) The repertoire of nonverbal behavior categories: origins, usage, and coding. Semiotica 1:49–98

Foucault M (1982) This is not a pipe. University of California Press, Los Angeles

Gabin B, Camerino O, Anguera M, Castañer M (2012) Lince: multiplatform sport analysis software. Procedia Soc Behav Sci 46:4692–4694

Gadamer HG (1980) Dialogue and dialectic: eight hermeneutical studies on plato. Yale University Press, New Haven, CT

Gallaher PE (1992) Individual differences in non-verbal behavior dimensions of style. J Pers Soc Psychol 63(1):133–145

Goffman E (1959) The presentation of self in everyday life. Doubleday and Company, New York

Goldin-Meadow S (2003) The resilience of language. Psychology Press, New York

Hall ET (1968) Proxemics. Current anthropology 9(2–3):83

Haynal-Reymond V, Jonsson GK, Magnusson MS (2005) Non-verbal communication in doctor-suicidal patient interview. In: Anolli L, Duncan S, Magnusson M, Riva G (eds) The hidden structure of social interaction, From Genomics to Culture Patterns. IOS Press, Amsterdam, pp 141–148

Heidegger M (1982) The basic problems of phenomenology. Indiana University Press, Bloomington

Izquierdo C, Anguera MT (2001) The role of the morphokinetic notational system in the observation of movement. In: Cavé C, Guatella I, Santi S (eds) Oralité et Gestualité. Interactions et comportements multimodaux dans la communication. L'Harmattan, Paris, pp 385–389

Jones SE, LeBaron CD (2002) Research on the relationship between verbal and nonverbal communication: emerging integrations. J. Commun. 52(3):499–521

Jonsson GK, Anguera MT, Blanco-Villaseñor A, Losada JL, Hernández-Mendo A, Ardá O, TCamerino, Castellano J (2006) Hidden patterns of play interaction in soccer using sof-coder. Behav Res Methods Instrum Comput 38(3):372–381

Jordan JS, Ghin M (2006) (photo-) consciousness as a contextually emergent property of self-sustaining systems. Mind Matter 4(1):45–68

Kendon A (1993) Space, time and gesture. Degrès 74:3–16

Kendon AE (1969) Nonverbal communication, interaction and gesture. Mouton, The Hague

Laban RV, Ullman L (1988) The mastery of movement. Northcote House, Plymouth, MA

Le Poire BA, Yoshimura SM (1999) The effects of expectancies and actual communication on non-verbal adaptation and communication outcomes: a test of interaction adaptation theory. Commun Monogr 66(1):1–30

Lemke J (2000) Across the scales of time: Artifacts, activities and meanings in ecosocial systems. Mind Cult Act 7(4):273–290

Magnusson MS (1996) Hidden real-time patterns in intra- and inter-individual behavior. Eur J Psychol Assess 12(2):112–123

Magnusson MS (2000) Discovering hidden time patterns in behavior: T-patterns and their detection. Behav Res Methods Instrum Comput 32(1):93–110

Magnusson MS (2005) Understanding social interaction: discovering hidden structure with model and algorithms. In: Anolli L Jr, SD, Magnusson MS, Riva G, (eds) The hidden structure of interaction: From neurons to culture patterns. IOS Press, Amsterdam, pp 3–22

McNeill D (2005) Gesture and thought. University of Chicago Press, Chicago

Merleau-Ponty M (1962) Phenomenology of perception. Routledge, London

Moore A (1996) College teacher immediacy and student ratings of instruction. Commun Educ 45(1):29–39

Nietzsche F (1954) Thus Spoke Zarathustra. Random House, New York

Pattern Vision (2001) THEME coder (software). http://www.patternvision.com, Retrieved 15 Jan 2002

Pozzer-Ardenghi L, Roth WM (2008) Catchments, growth points, and the iterability of signs in classroom communication. Semiotica 172:389–409

Rodriguez J, Plax TG, Kearney P (1996) Clarifying the relationship between teacher nonverbal immediacy and student cognitive learning: affective learning as the central causal mediator. Commun Educ 45:293–305

Roth WM (2001) Gestures: their role in teaching and learning. Rev Educ Res 71(3):365–392

Roth WM (2004) Gestures: the leading edge in literacy development. In: Saul W (ed) Border crossing: essays on literacy and science, International Reading Association & National Science Teachers Association, pp 48–70

Roth WM, Lawless D (2002) Scientific investigations, metaphorical gestures, and the emergence of abstract scientific concepts. Learn Instr 12:285–304

Sakaguchi K, Jonsson GK, Hasegawa T (2005) Initial interpersonal attraction and movement synchrony in mixed-sex dyads. In: Anolli L, Duncan S, Magnusson M, Riva G (eds) The hidden structure of social interaction. From Genomics to Culture Patterns. IOS Press, Amsterdam, pp 107–120

Streeck J, Jordan JS (2009) Communication as a dynamical self-sustaining system: the importance of time-scales and nested context. Commun Theory 19:445–464

Wiener M, Devoe S, Robinson S, Geller J (1972) Non-verbal behaviour and non-verbal communication. Psychol Rev 79(3):185–214

Wise J (2000) Home: territory and identity. Cult Stud 14:295–310

Wolfgang A (1997) Nonverbal behavior. Perspectives, applications and intercultural insights. Hogrefe & Huber Publishers., Seattle

Chapter 3
Is It Possible to Perform "Liquefying" Actions in Conversational Analysis? The Detection of Structures in Indirect Observations

M. Teresa Anguera

Abstract Conversational analysis allows one to study human interaction, which is of great interest because it is a spontaneous manifestation on the part of the participants. Communicative exchanges can be analyzed using a wide range of conceptual approaches and response levels. Various methodological decisions must be made, such as determining which dimensions—and, possibly, which subdimensions thereof—should be considered in the conceptual framework and what criteria should be applied for segmentation into units. The construction of an ad hoc indirect observation instrument is an especially important task, as it is the materialization of the researcher's specific interests in terms of bringing to light and/or prioritizing specific information. Each of the dimensions—or subdimensions thereof—permit the construction of behavior catalogues or category systems that then generate the respective codes. The application of the indirect observation instrument allows conversational episodes to be coded, usually by means of a computerized coding system that transforms conversational episodes into code matrices. The code matrices are the raw material used in whatever quantitative analyses are deemed appropriate by the researchers. The aim is usually to identify regularities that elucidate the underlying structure, any patterns that may emerge, and any vectors that may appear, and which in any case provide important information about associations between the codes representing the various behaviors or categories.

3.1 Introduction

A multitude of perspectives converge in social interaction. In keeping with Sidnell's (2013: 87) (Sidnell 2013: 87) that "a question remains as to *what* should be observed" these multiple perspectives must be adapted to the objectives of each study, and decisions made in this regard can have profound methodological consequences.

M. T. Anguera (✉)
Faculty of Psychology, Institute of Neurosciences, University of Barcelona,
Barcelona, Spain
e-mail: tanguera@ub.edu

© Springer Nature Switzerland AG 2020
L. Hunyadi and I. Szekrényes (eds.), *The Temporal Structure of Multimodal Communication*, Intelligent Systems Reference Library 164,
https://doi.org/10.1007/978-3-030-22895-8_3

In the present work, we will discuss the methodological analysis of conversational situations, approaching the subject from the perspective of *multimodal communication*, with a primary -but not exclusive- focus on speech among participants who do not follow a pre-established script and thus act spontaneously, following whatever rules govern their particular context.

According to Mondada (2013: 218), conversational analysis "looks at the endogenous organization of social activities in their ordinary settings: it considers social interaction as it is organized collectively by the co-participants, in a locally situated way, and as it is built incrementally through its temporal and sequential unfolding, by mobilizing a large range of vocal, verbal, visual and embodied resources, which are publicly displayed and monitored in situ." She adds: "Conversation analysis' naturalistic approach demands the study of 'naturally occurring activities' as they unfold in their ordinary social settings. This highlights the fact that for a detailed analysis of their relevant endogenous order, recordings of actual situated activities are necessary" (Mondada 2013: 218).

This stance should prompt us to reflect on what sorts of elements can converge in spontaneous conversation. It should be added that the approach taken in each case will have enormous consequences in terms of procedure. According to Selting (2013: 590), "participants in interaction use verbal, vocal and visual cues in co-occurrence and concurrence in order to organize their interaction." It should be mentioned, however, that the delimitation of response levels—a term proposed by Weick (1968) and discussed in Sect. 2.2.1 later on—is not always straightforward in scientific terms. Whereas verbal behavior that takes place in conversational interactions (Schegloff 2005) includes rhetorical, lexical-semantic, syntactic, phonetic-segmental and phonological aspects, vocal behavior includes aspects such as prosody and voice quality, given that the syntagmatic relationships between syllables are not determined by the structure of the words and phrases. Some authors, such as Kelly and Local (1989), have referred to these aspects as *communicative signals*. Still, the issue is not straightforward with regard to *visual cues* (Selting 2013), given that some authors use the expression *nonverbal communication*, which has been criticized for inducing bias. In any event, the delimitation and classification of the various communicative elements or cues is an analytical question, not a practical problem related to the participants in an interaction.

What is relevant is the multiplicity of levels that come into play in an interaction, either by intersecting with one another or as juxtaposed or parallel levels. As Sidnell (2006: 379f) notes: "Current work on multimodality focuses on questions of integration (or 'reassembly' as Schegloff 2005 put it) by putting at the forefront the question of how different modalities are integrated so as to form coherent courses of action." Let us consider Sidnell's proposal regarding this complex issue:

> To investigate multimodality, one needs to pay attention to the level of structured activities: those situated activity systems within which analysts and the coparticipants encounter gestures, directed gaze, and talk working together in a coordinated and differentiated way. This is a unit of interaction that is relatively discrete; has a beginning, middle, and end, and provides a structure of opportunities for participation. (Sidnell 2006: 380)

Various multimodal resources (gestures, gaze, head movements, facial expressions, posture, handling of objects, etc.) are mobilized during conversation, and many authors—such as Bressem (2013), Heath (1986, 1989), Poyatos (2002a, b), and Sidnell and Stevers (2013)—have shown interest in expanding the possibilities and increasing the rigor of research in this field.

Since the seminal works of Schegloff (1984), Goodwin (1981) and Heath (1986), the analysis of the elements that are co-present in an interaction and therefore co-participate in it has allowed us to identify and distinguish deliberate actions. In this regard, there has been convergence with other pioneering research on gestures, including that of Kendon (1990) and McNeill (1981), who argue that gestures and speech are not "modules" that function separately in communication. Gestures—including emblems, which have a strong symbolic or emblematic value—play an important role in conversational analysis.

Gestures have enormous linguistic potential (Müller 2013). They perform functions having to do with representation, signaling and expressing emotions, as well as pragmatic functions. Hence, gestures have frequently been considered in studies of social interaction as part of the body/language dichotomy (Bonaiuto and Maricchiolo 2013). As Müller (2003: 263) notes, "The gestures (…) are part of the narrative structure of a verbally and bodily described event. They create a visual display of the narrative structure and thus turn the story telling into a multi-modal event: something to listen to but also something to watch (…) they are natural elements of an everyday rhetoric of telling a story in conversation."

Even if we largely accept Sperber and Wilson (1986) view that communication is an exchange of representations, should be added that not all communication implies language. Whether spoken or written, language is one possible medium and form of representation, but it is not the only one—there are "body languages" that perform the same function. Hence, this body/language dichotomy is a common element of various paradigms that have been proposed as ways of understanding interpersonal communication (coding/decoding, intentionalism, dialogic paradigm, anticipatory perspective, etc.).

However, multimodal interaction does not consist exclusively of verbal and non-verbal actions (spoken language and gestures, posture, gaze, etc.); it also involves material objects in the environment (mobile telephone, keyring, toy, etc.) and even the setting itself (office, public park, store, etc.). Indeed, multimodal interaction appears to be a very intricate task, and as such it requires a flexible and rigorous methodological approach.

3.2 Towards an Optimal Methodology for Conversational Analysis

Our starting point is generic spontaneous multimodal communication, which takes place without any restrictions imposed by a script and can occur in any imaginable situation. We are interested in the analysis of conversations, which are a particular case of this larger category. The different response levels mentioned above, with verbal behavior as the primary area of focus, are structures that basically operate in parallel and co-occurrently in the construction of speaking turns in conversational interaction.

3.2.1 Systematic Observation

Observational methodology is the best option for studying this kind of reality. According to Sidnell (2013: 86), "Conversation Analysis begins with observation: listening to and, where the data is video-recorded, simultaneously watching a segment of talk-in-interaction typically aided by some form of transcription that can accommodate any observations one makes."

One possible methodology—indeed, the optimal one—is systematic observation, which guarantees a high degree of rigor and flexibility. This approach allows researchers to video- and audio-record perceptible behaviors and develop a scientific procedure for studying and analyzing spontaneous behaviors that occur in conversational episodes (Anguera 1979, 2003, 2017; Anguera and Izquierdo 2006; Portell et al. 2015; Sánchez-Algarra and Anguera 2013).

In many cases, researchers are interested in studying the profile of a particular type of conversation, such as doctor-patient dialogue and conversations in working groups, intra-family conversations, mediation in conflicts. This approach, if applied repeatedly—proceeding from the particular to the general, within a framework of inductive logic—can even lead to conceptual developments related to the particular conversation type being studied. Sidnell (2013: 86) writes: "We will be using observation as a basis for theorizing. Thus we start with things that are not currently imaginable, by showing that they happened."

We must begin by differentiating between direct and indirect observation. Direct observation uses visual perception, specifically, images captured on video. Video recordings are a good way to obtain information about gestures, posture, eye contact, facial expressions, etc. Indirect observation, which plays a crucial—though not exclusive—role in conversational analysis, focuses on verbal behavior. i.e., the content of the message.

Indirect observation is largely based on textual material generated either indirectly from transcriptions of audio recordings of verbal behavior in natural settings (e.g., conversation, group discussions) or directly from narratives (e.g., letters of complaint, tweets, forum posts). It may also feature seemingly unobtrusive objects that can provide relevant insights into daily routines. All these materials constitute an extremely rich source of information for studying conversation episodes, and they are continuously growing with the burgeoning of new technologies for data recording, dissemination and storage.

Indirect observation always yields qualitative data, but qualitative data have been widely criticized as unsystematic and lacking in objectivity. Narratives are an excellent vehicle for studying everyday life, but they need to be quantified and subsequently processed without loss of relevant information. Researchers therefore need a structured system for organizing and preparing the frequently heterogeneous information resulting from indirect observation for objective analysis (Anguera et al. 2018a, b).

A recent paper on indirect observation (Anguera et al. 2018b) states that this approach deals with three main types of materials—verbal materials, spoken texts and written texts—although they vary widely in terms of format and the ways in which the various types of documents are used. The use of complementary sources is very advisable (Darbyshire et al. 2005).

In most situations, the two types of observation—direct and indirect—should be used in a complementary manner, given that each type offers a different degree of perception (partial in the former case, total in the latter).

3.2.2 Basic Decisions in the Analysis of Turn-Taking

The wealth of information derived from a conversation episode must be systematized, precisely because our aim is to persue an exact scientific approach. This systematization requires that we take great care in how we make crucial decisions, specifically how we determine what dimensions will be used and how the conversation will be segmented into units.

3.2.2.1   Selection of Dimensions/Subdimensions

First, we must examine the various elements comprising the communicative situation—the participants, the activity being performed, and the physical, social, organizational, institutional, etc., characteristics of the context—which can provide information at any given time, and indeed at all times. This is what Norris (2013) called the "horizontal simultaneity" of actions. At any given time, each participant may be performing an activity, using the context in a certain way, and has a "life of his own" in the sense that he may be interacting with one or more participants simultaneously—or with none—either individually or collectively. Our aim is to adequately channel this reality as it happens, either in its entirety or by selecting only the elements that are of interest to us for the purposes of our study. In other words, communicative multimodality is a fact and it is sometimes unrealistic to attempt to grasp it in its totality. It is realistic, however, to choose particular elements that we consider relevant in each case. In systematic observation studies, we call this the selection of dimensions or criteria (which may, in turn, be divided into subdimensions). These are the elements that sustain the study conceptually, and on the basis of which the observational measurement instrument is constructed.

As the processing of any multimodal communication situation demonstrates, it is hard to study conversations of this sort. Using established delimitation criteria, we can assume that a speaking turn begins with the utterance of the first syntactic unit. The speaker may then go on to expand the syntactic constructions, producing deliberate sounds or mid-sentence pauses. Gestures also play a crucial role in the organization of speaking turns, as explained by Müller (2009):

> Gestures are part and parcel of the utterance and contribute semantic, syntactic and prag-matic information to the verbal part of the utterance whenever necessary. As a visuo-spatial medium, gestures are well suited to giving spatial, relational, shape, size and motion infor-mation, or enacting actions. Speech does what language is equipped for, such as establishing reference to absent entities, actions or events or establishing complex relations. In addition, gestures are widely employed to turn verbally implicit pragmatic and modal information into gesturally explicit information. (Müller 2009: 517)

Depending on the focus and theoretical framework of a particular study, one or more of the dimensions can be unfolded into subdimensions. For example, in a conversation in which verbal and gestural behavior is of interest—clearly a situation in which direct and indirect observation should be used in a complementary fashion— the gestural dimension can be unfolded into multiple subdimensions on the basis of topographic anatomic criteria (laterality, etc.) or other characteristics such as interactive components and qualifying features (Poyatos 1986).

In any case, the first decision is to delimit the dimensions or response levels (Weick 1968). Table 3.1 shows a simple example (without subdimensions) that includes the sequence, the participants, the transcription of verbal behavior and notes about the tone of voice. As a recent paper explained (Anguera 2017), the speech stream contains elements in general are not transcribed, such as intonation and pauses (Reynar 1998). In a study of the relationships between intonation and discourse structure, Hirschberg and Grosz (1992) found a correlation between intonational elements and the labeling of discourse parts. Hirschberg and Nakatani (1996) showed that considering speech in conjunction with transcriptions achieved greater consistency than the use of text alone and they found a high level of inter-labeler agreement in the segmentation of the discourse into units.

In any conversational situation—whether face-to-face, via videoconference, or through text messages—"we take turns to talk. Each turn we take is designed to 'do' something. Inter-action consists of the interplay between what one speaker is doing in a turn-at-talk and what the other did in their prior turn, and furthermore between what a speaker is doing in a current turn and what the other will do in response in his/her next turn" (Drew 2013: 131).

Turns are "macro" units conceived, in Hayashi's (2005: 47) words, as "multimodal packages for the production of action (and collaborative action) that make use of a range of different modalities, e.g., grammatical structure, sequential organization, organization of gaze and gesture, spatial-orientational frameworks, etc., in conjunc-tion with each other." Adopting the perspective of this view, which is situated in a propositive temporal framework, we consider conversation as consisting of speaking turns and fully agree with Hayashi's (2005: 47–48) conceptualization of a turn as "a temporally unfolding, interactively sustained domain of embodied action through which both the speaker and recipients build in concert with one another relevant actions that contribute to the further progression of the activity in progress."

Another important aspect to bear in mind is that the temporal perspective of conversational analysis leads to have two different approaches to the regulation of

Table 3.1 Fragment of a conversation taken from a mediation session [DDD and EEE are the parties in conflict; AAA and BBB are their respective mediators]

Context: The two mediators [before the start of session 1]

1	AAA:	We going to try to find a solution to the conflict	
2	BBB:	Yes, but...	
3	AAA:	What would you like to say?	
4	BBB:	That the conflict could evolve and aspects unknown to us might arise. We need to analyze all the information we have very carefully	
5	AAA:	It's not an easy case	
6		[Silence]	

Context: Session 1 between the parties in conflict (brother and sister) and the two mediators

7	AAA:	We're about to start the first of several sessions that we are going to have. We would like to ask each of the parties to speak calmly and politely during this conversation	
8	DDD:	I'll go first. The initial issue between my sister and me has to do with the ownership and use of a house. Our parents died in an accident and there was no will	
9	EEE:	Okay, there was no written will, but they had said many times that the house would be for their daughter, who spent more time with them	
10	DDD:	That's not true! They said it would be for whoever took better care of them. You spent more time with them, but not taking care of them, quite the opposite, making trouble for them	In a forceful tone
11	EEE:	You're a liar!!!!!!!!!!!!!! I'm not going to put up with this!	Shouting
12	AAA:	This conversation is to talk, not to insult anyone	Conciliating tone
13	EEE:	…I'm not going to let my brother say that I made trouble for my parents…	Shouting
14		[Moment of silence]	
15	DDD:	You've always been a liar and a troublemaker, and you've always wanted to get something out of everything. You've taken advantage of all the circumstances. You never took care of Mom and Dad. You constantly made them upset and were always asking for money. You never did anything for them. You only went to their house to play golf and to get money for whatever you fancied	In a forceful tone
16	EEE:	I'm going to fight you on this! You have no right to attack me to get the house. No, no, noooooooooooooooooooooooo!!!	Shouting very loudly
17	BBB:	We need to talk in a civilized manner. Let's start by having each of you describe what your lives were like when you still lived with your parents. You go first, EEE	Conciliating tone

[End of initial fragment from session 1]

interactive episodes (Clift 2016). The first is *self-regulation* of speaking turns, in which each speaker tries to ensure that his speech does not overlap with anyone else's (i.e., "one speaker at a time") in the course of the interactional traffic. The second and clearly differentiated form of regulation is *turn-taking*. This approach is used in certain types of conversations (a focus group would be an extreme example), in which each participant "waits his or her turn," "has the floor," "takes a pass," etc. In this sort of multimodal interaction, "the use of pointing gestures predicting turn completions and projecting the emergence of possible next speakers" (Mondada 2007: 194) plays a key role in the organization of speaking turns and it has an important modulating effect. In any case, for a speaker in an interactive episode, such as a focus group, the prospect of an imminent turn acts as a constraint, and we must take this into account when we make decisions regarding the empirical aspect of a study.

3.2.2.2 Segmentation of the Conversation into Units

The second decision that must be made, according to Norris (2013), is how to segment the conversation into units. This is not an easy decision, as it depends on the "degree of zoom" we wish to apply in the multimodal communication analysis and it must therefore be considered carefully. Let us consider an imaginary episode—for example, a get-together between friends. A case could be made for the unit being the episode itself (i.e., the entire get-together). Other possibilities would be to treat speaking turns (interlocutory perspective), gestures, gazes and so onas units. Moreover, given the co-occurrence of the various dimensions and subdimensions, each one of these elements would require an appropriate type of unit. In the example mentioned above—a gathering of several friends—eye contact between two people could be one unit, which would be nested within another unit, like a side dialogue between two members of the group. As Norris (2013) writes, "The essence in multimodal (inter)action analysis is always the linking of the multitude of modes that are used in (inter)action when social actors act and communicate" (Norris 2013: 281).

For this second decision, we propose a hierarchical structure resembling the various levels of a pyramid, either bottom-up, with the most molecularized units at the bottom and the levels becoming more molarized as you go up, or top-down, where we start with a series of conversational episodes that become progressively molecularized. The molecularity-molarity range of a particular study will obviously be linked to the objectives considered in each case.

Dickman (1963) and Birdwhistell (1970) proposed that the behavioral flow should be segmented into units. This approach was later developed through the so-called "rule of the three Ds" (Anguera and Izquierdo 2006): *delimitation* (the beginning and end of each unit must be specified), *denomination* (each unit must have a name) and *definition* (each unit must be definable). It is possible to establish a gradation of interactive units as a function of their "granularization" (a term adapted from Schegloff [2000] and applied at the methodological level). Various levels of granularity can be established (Schegloff and Sacks 1973: 325) as a function of the possibilities that exist on a continuum ranging from most molar to most molecular (Anguera 2017).

It should be mentioned that the creation of the code matrix itself presents certain difficulties. These difficulties increase when, in successive turns, each of the various response levels contributes different information, thus requiring the multimodal integration of interactive sequences, which Schegloff (2005) described using the curious expression "talk and other conduct in interaction." Sometimes—often, in fact—the segmentation of the various response levels or dimensions into units may not be coincident. In these cases, it is advisable to use the response level that is most appropiate to the study's objectives priority in the segmentation process.

3.3 Methodological Process: Transcription, Use of Observation Instrument, Coding and Data-Quality Control

Systematic observation applied to conversational analysis must conform to the established procedure. Here, however, we are interested in discussing the specific aspects of this technique that arise from the fact that it combines direct and indirect observation, with indirect observation playing the predominant role.

3.3.1 Transcription

Collecting the information generated in a conversation is a key aspect of conversational analysis. This information is transformed into a dataset through the intermediate operation of transcription and it is then used as the raw material in the subsequent analysis. It is worth remembering, as Hepburn and Bolden (2013: 57) remark, that "transcripts are necessarily selective in the details that are represented."

Transcriptions are representations of communicative events that allow us to preserve the data—the spoken behavior—as it is produced. With the exponential technological advances of recent years, it has become possible to transcribe more and more types of data. As Bohle (2013) explains, the fundamental content is as follows:

(a) Selection of what to transcribe, with the appropriate contextual information.
(b) Segmentation of the observable behavioral flow into significant units.
(c) Structuring or arranging of the text—that is, the speaking turns, with the accompanying modulations of contextual information.
(d) Coding.

The decision regarding what to transcribe is very important, as it must be taken into account in the construction of the observation instrument.

Decisions about content are influenced by the theoretical framework adopted, the preconceptions of those doing the transcribing, the aim of the study and methodological considerations (Cook 1990; Ochs 1979).

As O'Connell and Kowal (2000) and Bohle (2013) rightly noted, transcripts are sources of tertiary data (either video or audio recordings). New technologies have made it possible to achieve high standards of accuracy, bringing about an "interest in interactional details that would have been overlooked in the past" (Duranti 1997: 123).

Nevertheless, as Cook (1990) remarks, this sensitive material is not free of problems:

> The problems posed for transcription by the introduction of context are of identity, quality and quantity. The first problem is to find a means of distinguishing relevant features, the second of devising a transcription system that is capable of expressing them, the third that even if such a system could be devised it would make the presentation of data (which in its actual production had occupied a short space or time) take pages of transcript […] (Cook 1990: 4).

Very few researchers have argued in favor of absolutely neutral transcripts without pre-selection. Our view is that there may be some degree of correspondence between the way in which transcription is carried out and the inductive or deductive nature of the study.

Lévi-Strauss (1963: 280) provides a clear example of a theoretically neutral position: "On the observational level, the main—one could almost say the only—rule is that all the facts should be carefully observed and described, without allowing any theoretical preconception to decide whether some are more important than others." Taking a similar stance and an inductive position, Scheflen (1966: 270) proposes: "We do not decide beforehand what is trivial, what is redundant, or what alters the system. This is a *result* of the research."

If the researcher is so inclined, this could lead to successive stages of research. In other words, an initial exploratory study could lead to a follow-up study to confirm the findings.

It is clear, however, that it is impossible to observe the entirety of a conversation. Each "molecule" of reality contains a data source, and we undoubtedly must establish filters. As Bohle (2013: 995) notes, "transcripts are inherently and unavoidably selective." However, the decision chains that lead to this point are very diverse, ranging from intuition to theoretical framework, research interests, methodological approach, etc., and, as Ochs (1979) notes, the specific objectives of a given study.

3.3.2 Use of the Observation Instrument

From Sidnell's (2013: 78) affirmation—"I then consider the importance of observation as a means for identifying a collectable, researchable phenomenon or practice"—it seems obvious that each chunk of conversational reality expands polyhedrally, exhibiting an unusual wealth of data, and that we must establish the appropriate filters and channels to eliminate redundancies and unwanted or irrelevant information, with the aim of striking a balance between the profusion of data that can be filmed or recorded, the filtering of this data, and the necessary systematization in the form of code matrices.

To implement this methodological approach in conversational studies, we must clearly define the object of study. Sidnell (2013) makes the case forcefully: "One basic goal of Conversation Analysis is to identify the actions that participants in interaction do and describe the particular practices of conduct that they use to accomplish them. While the meaning of **actions** here may be relatively transparent—i.e. asking, telling, requesting, offering, inviting, complaining, announcing and so on—the meaning of practices is likely less so" (Sidnell 2013: 78). Likewise, Heritage (2011: 212) states:

> A 'practice' is any feature of the design of a turn in a sequence that (i) has a distinctive character, (ii) has specific locations within a turn or sequence, and (iii) is distinctive in its consequences for the nature or the meaning of the action that the turn implements.

Both Sidnell and Heritage are telling us, very subtly, that we must in some way specify what we will register and how. The observation instrument—which is ad hoc, not standard—plays a key role in this regard, despite the fact that Gnisci, Maricchiolo and Bonaiuto (2013) consider that the roots of systemic observation are found in Stevens (1946) theory of scales of measurement.

The main function of the observation instrument is to make it possible for each unit of behavior in any dimension or subdimension to be assigned a code that has a meaning. In Sect. 2.2.1 above, we discussed the decision regarding the specification of dimensions considered relevant in each case, as well as their disaggregation into subdimensions. With the support of the theoretical framework, if there is one, and the expertise of researchers and professionals, there are two possibilities (Anguera 2003; Anguera et al. 2007; Sánchez-Algarra and Anguera 2013):

(a) The field format instrument: A catalogue of behaviors, or a list of names of behaviors, with the corresponding codes, which must meet the criterion of mutual exclusivity, is created on the basis of each of the dimensions (or, where appropriate, the subdimensions thereof). The list is open and new items can be added indefinitely.
(b) The field format instrument combined with category systems: A category system is created on the basis of one, several or all dimensions (or, where appropriate, the subdimensions thereof), with corresponding definitions and codes, and the conditions of exhaustiveness and mutual exclusivity must be met. The list is closed and cannot be modified. If only some of the dimensions/subdimensions permit the creation of category systems, behavior catalogues are created for the others, as in the field format instrument. The lists or these catalogues are open.

Field format instruments, either alone or in combination with category systems, are proving to be very useful in studies of communication in various contexts (Arias-Pujol and Anguera 2017; Castañer et al. 2013; García-Fariña et al. 2016, 2018). Importantly, each dimension/subdimension has a system of codes (behaviors in some

Table 3.2 Code matrix with progressive disaggregation into subdimensions. Each box contains the assigned code, which corresponds to each of the observation units (directly observed behaviors) or textual units (text fragments in indirect observation) represented by the given row, showing the corresponding criterion or subcriterion of the observation instrument, in the given column. Each unit can be coded using one, several or all criteria, and some can remain empty. [Adapted from Anguera (2017)]

		Dimensions and subdimensions (i)												
		D1				D2				..	Di			
		D11		D12		D21		D22		..	Di1		Di2	
		D111	D112	D121	D122	D211	D212	D221	D222	..	Di11	Di12	Di21	Di22
Observation units or textual units (n)	U1													
	U2													
	...													
	Un													

cases and categories in others) that allow each unit of behavior of each dimension or subdimension to be assigned the corresponding code. The structure of this is shown in Table 3.2.

3.3.3 Coding

Once the observation instrument has been constructed, each behavior must be assigned the corresponding code. These codes have various functions (Burgoon et al. 2013) and can be disaggregated into simpler behaviors or aggregated to represent more complex behaviors. In our approach (Anguera 2017; Anguera et al. 2017b), chains are formed as a function of co-occurrence, conforming the rows of the code matrices created through the coding of the conversations.

Of course, these matrices are rarely regular. Their profile is usually irregular, since we cannot expect that each one of the successive units will have a code corresponding to each of the dimensions and/or subdimensions thereof. This is shown graphically in Fig. 3.1.

Coding can be performed either manually or—as is done in practically all situations nowadays—with computers. Code, according to Burgoon, Guerrero and Floyd (2010: 19), is "a set of signals that is usually transmitted via one particular medium or channel." A wide range of computer programs are available for this purpose: ATLAS.ti, AQUAD, NUDIST, NVIVO, etc. Nearly all such programs are free, although some commercial options are widely used. Some, such as NEU-ROGES+ELAN, have successfully combined the observation instrument with the registration tool (Lausberg and Sloetjes 2016).

As we have seen in a few recent studies (Anguera 2017), one excellent new option, which we have used, is to apply the codes that appear in the right-hand pane of ATLAS.ti, as shown in Fig. 3.2. These codes are transformed, using a minimal number

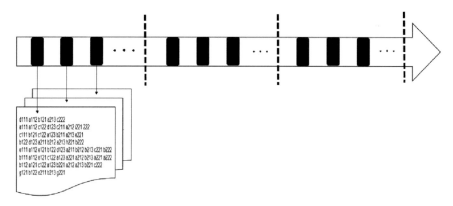

Fig. 3.1 Code matrices corresponding to each conversational episode over different time periods [Adapted from Anguera (2017)]

Fig. 3.2 Screenshot from the ATLAS.ti program highlighting the right-hand pane, which shows the corresponding codes. Data coming from García-Fariña (2015: 253). [Modified from Anguera et al. (2017b)]

of manual operations, into a code matrix (usually irregular), in which the minimum number of codes is one and the maximum number is the number of dimensions in the observation instrument. The codes are then ready for processing in THEME.

In accordance with the selected approach and following the example shown in Table 3.2, the data values, in the form of a code matrix, are entered in one of n conversation sessions in a hypothetical study (Fig. 3.1). Of course, each session will generate a code matrix in the dataset and each row of the code matrix will show the co-occurrences (corresponding to the various subcriteria) that have taken place in each of the successive units (behavioral or textual). The code matrix in general will not be regular.

3.3.4 Data Quality Control

After collecting the data, the quality of the data must be verified. Observational methodology undeniably requires an effort to detect registration biases and short-comings in inter-observer agreement in the assignment of codes in the observation instrument to the respective units delimited in the dimensions/subdimensions.

Even greater caution is required in indirect observation because the risk of inference is greater. At least three data points are required to calculate the statistical agreement.

Numerous coefficients exist for quantitatively verifying the quality of data in a wide range of situations. One widely used measure in indirect observation is Krippendorff's canonical agreement coefficient (Krippendorff 2013), which is an adaptation of Cohen's kappa coefficient (Cohen 1960) for analyzing three or more datasets. It can be calculated in the free HOISAN software package [www.menpas. com] (Hernández-Mendo et al. 2012).

A more qualitative method, the consensus agreement method (Anguera 1990), is gaining increasing recognition in indirect observation and other studies. In this method, at least three observers work together to discuss and agree on the most suitable code for each unit from the observation unit (Anguera et al. 2018a, b). This method has obvious advantages, as it produces a single dataset and frequently results in a better observation instrument thanks to the detection of possible gaps and shortcomings. While it offers significant guarantees of quality, it also carries risks. An observer may defer to the decisions of a more senior or "expert" colleague, for example, and the need to agree can also give rise to frictions or conflicts. The results of the consensus agreement method can be complemented by quantitative measures of agreement (Arana et al. 2016).

3.4 T-Pattern Detection in Conversational Analysis

Once the data quality control has been completed, the dataset is analyzed in order to determine the structure that is not visible to the naked eye, and to identify regularities that govern conversational episodes, as Sidnell (2013: 77) explains perfectly: "Conversation Analysis is meant to be a kind of exploration, the goal of which is the discovery of previously unknown regularities of human interaction."

Conversations are organized sequentially. Each sentence follows the one that precedes it, forming the *adjacency pairs*—to use the term coined by Schegloff and Sacks (1973)—that make up the most basic sequence. This sequence can be expanded in various ways, some very typical, such as the classic *"How are you?"* Very often, however, expansions occur in the sequences as a function of decisions made by the conversation's participants. An impressive interplay of possibilities make up the reality of each conversation, including interspersed pauses, sudden changes of topic and tone, thematic evolution, and the addition of new participants (either real or virtual, as in the case of a mobile telephone call or text message received).

Diachrony, or sequentiality, has been recognized as a basic attribute of interactive processes since the second half of the 20th century (Heyns and Lippitt 1954). As noted by Gnisci, Bakeman and Maricchiolo (2013), the term "sequential" allows us to characterize the dynamic aspect of an interaction. Participants in a conversation must temporally coordinate various aspects of their own behavior (mainly speech and gestures, as well as tone of voice) with the behavior of the other people taking turns participating in the conversation.

Qualitative approaches to the study of interaction from a sequential point of view have been proposed (Sacks 1972; Sacks et al. 1974). However, my colleagues and I use a different approach: we envisage a mixed-method perspective (Anguera 2017; Anguera et al. 2018a; Sánchez-Algarra and Anguera 2013) based on the quantitative treatment of qualitative data that are obtained through systematic observation and are properly organized.

Besides the diachronic perspective, another important pillar—one which echoes the title of this book—must be preserved: the *temporal structure of multimodal communication*. According to Hepburn and Bolden (2013: 59), "One of the most consequential contributions of CA [conversational analysis] to the study of talk and social action was the introduction of timing and sequential position in interaction. CA research has amply demonstrated the incredible precision with which interlocutors coordinate their talk," not to mention their speech overlaps, latching, gaps, pauses, etc., as well as features that accompany speech such as inhalation, laughter, voice-raising and shouting. Moreover, as indicated in Sect. 2.2.1., the various response levels selected for each study (eye contact, gestures, body movements, etc.) have also been incorporated.

As my colleagues and I explained in a recent paper (Anguera et al. 2017a), systematic coding allows the complexity of human communication to be broken down into a series of episodes or strings of episodes represented by codes. These codes form code matrices perfectly aligned with the syntactic rules of THEME [www.patternvision.com]. THEME is a free data analysis software tool first developed by Magnusson (1988, 1989, 1993, 1996, 2000, 2005, 2006, 2016) some thirty years ago and continuously developed up to the current version (THEME v. 6 Edu). It detects temporal patterns, or T-patterns, which are structures that show the temporal relationships between different codes (elements of the communication process) according to their sequence and separation in time (distance). These structures are imperceptible to the naked eye, but can be extracted from the code matrices through a robust quantitative analysis (Anolli et al. 2005).

The assumption underlying the T-pattern detection method is that complex types of human behavior, and also conversation episodes, have a temporal structure that cannot be fully detected through traditional observational methods or mere quantitative statistical logic. T-patterns emerge as a result of a mathematical process that is automated in the form of an algorithm in THEME. The T-pattern detection method can identify structural analogies across very different levels of organization and permit an important shift from quantitative to structural analysis.

T-pattern detection studies have been conducted in a diverse range of scientific domains and also in multimodal communication situations such as conversational analysis (Baraud et al. 2016; Blanchet et al. 2005; García-Fariña 2015; Vaimberg 2010) or phonological awareness (Suárez et al. 2018).

In conversational analysis, the duration of verbal behavior units (indirect observation) is often considered irrelevant and therefore is not recorded, whereas the duration of gestural, postural, etc., behavior (direct observation) characteristic of multimodal communication usually is recorded. In order to resolve these situations and detect T-patterns, Iglesias et al. (2016) used the THEME v. 6 Edu software program and assigned a constant duration ($= 1$) to each event-type (Lapresa and Alsasua et al. 2013), as what was important in their analysis was not the duration of each phrase or the distance between phrases (which was very similar) but rather their internal sequentiality. To illustrate this point, using the observation instrument shown in Fig. 3.3 and the THEME software tool, 5541 T-patterns were detected; the first such T-pattern being shown in Fig. 3.4.

Given the vastness of the literature, it is not possible to mention every possible situation that can arise, but we can go over the basics of transforming the transcription of conversational episodes into a code matrix that can be used to detect seemingly invisible structures and study how they evolve. Interesting situations include those related to the internal symmetry of T-patterns (Anguera 2005), the progressive degradation of T-patterns from start to finish (Lapresa et al. 2015), qualitative and

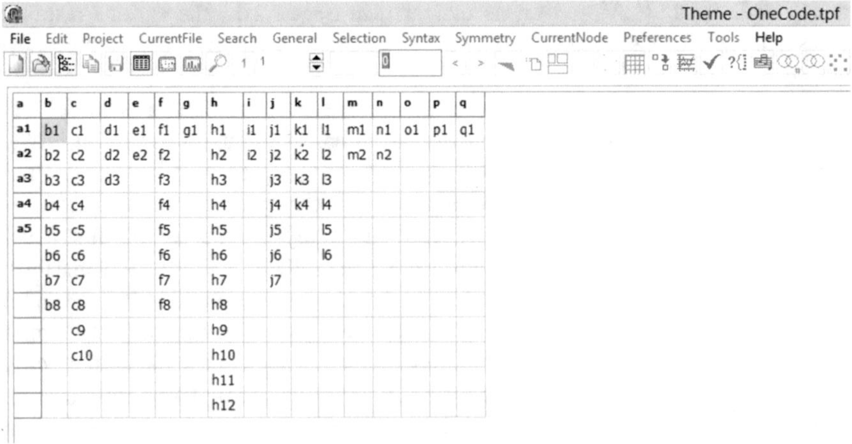

Fig. 3.3 Observation instrument entered into the THEME program. The dimensions are as follows: a, b, c, d, e, f, g, h, i, j, k, l, m, n, o, p, q. The respective columns show the behavior catalogues/category systems corresponding to each dimension

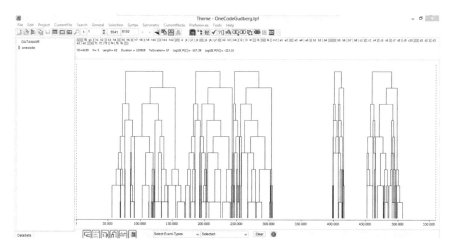

Fig. 3.4 T-pattern 1 (out of a total of 5541) corresponding to data values recorded by the observation tool shown in Fig. 3.3 The parameters considered are as follows: $p < 0.00005$; minimum number of occurrences = 5

quantitative filters for selecting and interpreting T-patterns (Amatria et al. 2017), and T-pattern analysis as a multivariate approach to the detection of temporal behavior structures (Casarrubea et al. 2015).

In many cases, the interruption of the sequence as a result of decisions by one or more participants or external eventualities leads to the generation of incomplete patterns, which permit a quantitative analysis based on qualitative data that provides results which present reality as it is.

3.5 What Do "LIQUEFYING" Actions in Conversational Analysis Involve?

Having seen that it is possible to start from a conversational situation and make further use of it, let us consider the strengths and weaknesses of this approach.

The strengths have to do with the fundamental characteristic of mixed methods (Anguera et al. 2018a, b) and the application of this approach to the development of methodology for indirect observation.

In addition to the widespread use of mixed methods to complement qualitative and quantitative elements, this approach focuses on one of the three integration operations—merge, connect and nest—proposed by various authors (Creswell and Plano Clark 2011). The operation in question is "connect," which we achieve by transforming qualitative data while preserving their integrity in all aspects of interest to us.

The essential elements of the process were described, including the construction of an indirect observation instrument, transcription, registration and coding. The aim of this procedure is to convert a chunk of reality—that is, a recorded conversation—into a code matrix. Although the information in this code matrix remains qualitative, it is organized, arranged in a systematized manner, and prepared on the basis of a parameter of order or sequence—that is, by obtaining information diachronically, over time, at the intra-session level (from the beginning to the end of a session or conversational episode) as well as the inter-session level (conversational episodes taking place over multiple sessions and maintaining a thematic or strategic homogeneity).

The transformation of the data into code matrices enabled us to carry out some of the various options that are available for the multivariate quantitative analysis of categorical data. The extensive literature on this subject, as well as our own experiences, indicate that T-pattern detection plays an essential role, although complementary analytical techniques such as polar coordinate analysis should also be considered.

Acknowledgements The author gratefully acknowledges the support of a Spanish government subproject *Integration ways between qualitative and quantitative data, multiple case development, and synthesis review as main axis for an innovative future in physical activity and sports research* [Grant number PGC2018-098742-B-C31] (Ministerio de Ciencia, Innovación y Universidades, Programa Estatal de Generación de Conocimiento y Fortalecimiento Científico y Tecnológico del Sistema I+D+i), that is part of the coordinated project *New approach of research in physical activity and sport from mixed methods perspective* (NARPAS_MM) [SPGC201800X098742CV0]. In addition, the author is grateful for the support of the Generalitat de Catalunya research group GRUP DE RECERCA I INNOVACIÓ EN DISSENYS (GRID). Tecnología i aplicació multimedia i digital als dissenys observacionals [Grant number 2017 SGR 1405].

References

Amatria M, Lapresa D, Arana J, Anguera MT, Jonsson GK (2017) Detection and selection of behavioral patterns using theme: A concrete example in grassroots soccer. Sports 5:20. https://doi.org/10.3390/sports5010020

Anguera MT (1979) Observational typology. Qual Quant Eur-Am J Methodol 13(6):449–484

Anguera MT (1990) Metodología observacional [Observational methodology]. In: Arnau J, Anguera MT, Gómez J (eds) Metodología de la investigación en Ciencias del Comportamiento. Universidad de Murcia, Murcia, pp 125–238

Anguera MT (2003) Observational methods (general). In: Fernández-Ballesteros R (ed) Encyclopedia of psychological assessment, vol 2. Sage, London, pp 632–637

Anguera MT (2005) Microanalysis of t-patterns. analysis of symmetry/asymmetry in social interaction. In: Anolli L, Duncan S, Magnusson M, Riva G (eds) The hidden structure of social interaction. From Genomics to Culture Patterns., IOS Press, Amsterdam, pp 51–70

Anguera MT (2017) Transiciones interactivas a lo largo de un proceso de desarrollo: Complementariedad de análisis [Interactive transitions throughout a development process: Complementarity of analysis]. In: Santoyo C (ed) Mecanismos básicos de toma de decisiones: Perspectivas desde las ciencias del comportamiento y del desarrollo, CONACYT 178383/UNAM, Mexico City, pp 179–213

Anguera MT (in press) Desarrollando la observación indirecta: Alcance, proceso, y habilidades metodológicas en el análisis de textos [Development of indirect observation: Scope, process and methodological abilities in textual analysis]. In: Santoyo C (ed) Análisis de patrones de habilidades metodológicas y conceptuales de análisis, planeación, evaluación e intervención, UNAM/PAPIIT, IN306715, Mexico City

Anguera MT, Izquierdo C (2006) Methodological approaches in human communication: From complexity of perceived situation to data analysis. In: Riva G, Anguera MT, Wiederhold BK, Mantovani F (eds) From communication to presence. Emotions and culture towards the ultimate communicative experience, IOS Press, Amsterdam, Cognition, pp 203–222

Anguera MT, Magnusson MS, Jonsson GK (2007) Instrumentos no estándar [Non standard instruments]. Avances en Medición 5(1):63–82

Anguera MT, Camerino O, Castañer M, Sánchez-Algarra P, Onwuegbuzie AJ (2017a) The specificity of observational studies in physical activity and sports sciences: Moving forward in mixed methods research and proposals for achieving quantitative and qualitative symmetry. Front Psychol 8:2196. https://doi.org/10.3389/fpsyg.2017.02196

Anguera MT, Jonsson GK, Sánchez-Algarra P (2017b) Liquefying text from human communication processes: A methodological proposal based on t-pattern detection. J Multimodal Commun Stud 4(1–2):10–15

Anguera MT, Blanco-Villaseñor A, Losada JL, Sánchez-Algarra P, Onwuegbuzie AJ (2018a) Revisiting the difference between mixed methods and multimethods: Is it all in the name? Qual Quant. https://doi.org/10.1007/s11135-018-0700-2

Anguera MT, Portell M, Chacón-Moscoso S, Sanduvete-Chaves S (2018b) Indirect observation in everyday contexts: Concepts and methodological guidelines within a mixed methods framework. Front Psychol 9:13. https://doi.org/10.3389/fpsyg.2018.00013

Anolli L, Duncan S, Magnusson MS, Riva G (eds) (2005) The Hidden Structure of Interaction. From neurons to culture patterns. IOS Press, Amsterdam

Arana J, Lapresa D, Anguera MT, Garzón B (2016) Ad hoc procedure for optimising agreement between observational records. Anales de Psicología 32(2):589–595

Arias-Pujol E, Anguera MT (2017) Observation of interactions in adolescent group therapy: A mixed methods study. Front Psychol 8:1188. https://doi.org/10.3389/fpsyg.2017.01188

Baraud I, Deputte BL, Pierre JS, Blois-Heulin C (2016) Informative value of vocalizations during multimodal interactions in red-capped mangabeys. In: Magnusson MS, Burgoon JK, Casarrubea M, McNeill D (eds) Discovering hidden temporal patterns in behavior and interactions: T-Pattern detection and analysis with THEME. Springer, New York, pp 255–278

Birdwhistell RL (1970) Kinesics and context: Essays and body motion communication. University of Pennsylvania Press, Philadelphia

Blanchet A, Batt M, Trognon A, Masse L (2005) Language and behaviour patterns in a therapeutic interaction sequence. In: Anolli L, Duncan S, Magnusson M, Riva G (eds) The hidden structure of social interaction. From genomics to culture patterns. IOS Press, Amsterdam, pp 124–140

Bohle U (2013) Approaching notation, coding, and analysis from a conversational analysis point of view. In: Müller C, Cienki A, Fricke E, Ladewig S, McNeill D, Teßendorf S (eds) Body–Language–Communication. An International Handbook on Multimodality in Human Interaction, vol. 1, De Gruyter, Frankfurt, pp 992–1007

Bonaiuto M, Maricchiolo F (2013) Social psychology: Body and language in social interaction. In: Müller C, Cienki A, Fricke E, Ladewig S, McNeill D, Teßendorf S (eds) Body – Language – Communication. An International Handbook on Multimodality in Human Interaction, vol. 1, De Gruyter, Frankfurt, pp 258–275

Bressem J (2013) 20th century: Empirical research of body, language, and communication. In: Müller C, Cienki A, Fricke E, Ladewig S, McNeill D, Teßendorf S (eds) Body–Language–Communication. An International Handbook on Multimodality in Human Interaction, vol. 1, De Gruyter, Frankfurt, pp 393–416

Burgoon J, Guerrero LK, Floyd K (2010) Nonverbal communication. Allyn & Bacon, Boston

Burgoon JK, Guerrero LK, White CH (2013) The codes and functions of nonverbal communication. In: Müller C, Cienki A, Fricke E, Ladewig S, McNeill D, Teßendorf S (eds) Body–Language–Communication. An International Handbook on Multimodality in Human Interaction, vol. 1, De Gruyter, Frankfurt, pp 609–626

Casarrubea M, Jonsson GK, Faulisi F, Sorbera F, Di Giovannim G, Benigno A, Crescimanno G, Magnusson MS (2015) T-pattern analysis for the study of temporal structure of animal and human behavior: A comprehensive review. J Neurosci Methods 239:44–46

Castañer M, Camerino O, Anguera MT, Jonsson GK (2013) Kinesics and proxemics communication of expert and novice PE teachers. Qual Quant 47(4):1813–1829

Clift R (2016) Conversation analysis. Cambridge University Press, Cambridge

Cohen J (1960) A coefficient of agreement for nominal scales. Educ Psychol Meas 20:37–46

Cook G (1990) Transcribing infinity. problems of context presentation. J Pragmat 14:1–24

Creswell JW, Plano Clark VL (2011) Designing and conducting mixed methods research, 2nd edn. Saga, Thousand Oaks

Darbyshire P, MacDougall C, Schiller W (2005) Multiple methods in qualitative research with children: More insight of just more? Qual Res 5:417–436

Dickman HR (1963) The perception of behavioral units. In: Barker R (ed) The stream of behavior. Appleton-Century-Crofts, New York, pp 23–41

Drew P (2013) Turn design. In: Sidnell J, Stivers T (eds) The Handbook of conversation analysis. Wiley-Blackwell, New York, pp 131–149

Duranti A (1997) Transcription: From writing to digitized images. In: Duranti A (ed) Linguistic anthropology. Cambridge University Press, Cambridge, pp 122–161

García-Fariña A (2015) Análisis del discurso docente como recurso metodológico del profesorado de Educación Física en la etapa de Educación Primaria [Analysis of teacher-led discourse as a methodological resource for primary school physical education teachers]. Unpublised Doctoral Dissertation. Universidad de La Laguna, Tenerife, Spain

García-Fariña A, Jiménez Jiménez F, Anguera MT (2016) Análisis observacional del discurso docente del profesorado de educación física a través de patrones comunicativos [Observational analysis of teaching discourse physical education training teachers through communicative patterns]. Cuadernos de Psicología de Deporte 16(1):171–182

García-Fariña A, Jiménez Jiménez F, Anguera MT (2018) Observation of physical education teachers' communication: Detecting patterns in verbal behavior. Front Psychol 9:334. https://doi.org/10.3389/fpsyg.2018.00334

Gnisci A, Bakeman R, Maricchiolo F (2013) Sequential notation and analysis for bodily forms of communication. In: Müller C, Cienki A, Fricke E, Ladewig S, McNeill D, Teßendorf S (eds) Body–Language–Communication. An International Handbook on Multimodality in Human Interaction, vol. 1, De Gruyter, Frankfurt, pp 892–903

Gnisci A, Maricchiolo F, Bonaiuto M (2013b) Reliability and validity of coding systems for bodily forms of communication. In: Müller C, Cienki A, Fricke E, Ladewig S, McNeill D, Teßendorf S (eds) Body–Language–Communication. An International Handbook on Multimodality in Human Interaction, vol. 1, De Gruyter, Frankfurt, pp 879–892

Goodwin C (1981) Conversational organization: Interaction between speakers and hearers. Academic, New York

Hayashi M (2005) Joint turn construction through language and the body: Notes on embodiment on coordinated participation in situated activities. Semiotica 156(1/4):21–53

Heath C (1986) Body movement and speech in medical interaction. Cambridge University Press, Cambridge

Heath C (1989) Pain talk: The expression of suffering in the medical consultation. Soc Psychol Q 52(2):113–125

Hepburn A, Bolden GB (2013) The conversation analytic approach to transcription. In: Sidnell J, Stivers T (eds) The handbook of conversation analysis. Wiley-Blackwell, New York, pp 57–76

Heritage J (2011) Conversation analysis: Practices and methods. In: Silverman D (ed) Qualitative research: Theory, method and practice, 3rd edn. Sage, London

Hernández-Mendo A, López-López JA, Castellano J, Morales-Sánchez V, Pastrana JL (2012) Hoisan 1.2: Programa informático para uso en metodología observacional [Hoisan 1.2: Software for observational methodology]. Cuadernos de Psicología del Deporte 12:55–78

Heyns RW, Lippitt R (1954) Systematic observational techniques. In: Lindzey G (ed) Handbook of social psychology: I. Theory and method. Addison-Wesley, Oxford

Hirschberg J, Grosz B (1992) Intonational features of local and global discourse. In: Proceedings of the workshop on spoken language systems, DARPA, pp 441–446

Hirschberg J, Nakatani CH (1996) A prosodic analysis of discourse segments in direction-giving monologues. In: Proceedings of the 34th annual meeting of the association for computational linguistics, Santa Cruz, California, pp 286–293

Iglesias X, Tarragó R, Lapresa D, Anguera MT (2016) A complementary study of elite fencing tactics using lag sequential, polar coordinate, and t-pattern analyses. In: LaCOSA II. International conference on sequence analysis and related methods, polar coordinate, Lausanne, Switzerland

Kelly J, Local J (1989) Doing phonology. Observing, recording, interpreting. Manchester University Press, Manchester

Kendon A (1990) Conducting interaction: Patterns of behavior in focused encounters. Cambridge University Press, Cambridge

Krippendorff K (2013) Content analysis. An introduction to its methodology, 3rd edn. Sage, Thousand Oaks

Lapresa D, Camerino O, Cabedo J, Anguera MT, Jonsson GK, Arana X (2015) Degradación de t-patterns en estudios observacionales: Un estudio sobre la eficacia en el ataque de fútbol sala [Degradation of t-patterns in observational studies: A study on the effectiveness in futsal]. Cuadernos de Psicología del Deporte 15(1):71–82

Lausberg H, Sloetjes H (2016) The revised neuroges-elan system: An objective and reliable interdisciplinary analysis tool for nonverbal behavior and gesture. Behav Res Methods 48:973–993

Lévi-Strauss C (1963) Struct Anthropol. Basic Books, New York

Magnusson MS (1988) Le temps et les patterns syntaxiques du comportement humain: Modèle, méthode et programme theme. Revue des Conditions de Travail, hors série pp 284–314

Magnusson MS (1989) Structure syntaxique et rythmes comportementaux: Sur la detection de rythmes caches. Sci Tech Anim Lab 14(2):143–147

Magnusson MS (1993) THEME user's manual: With notes on theory, model and pattern detection method. University of Iceland, Reykjavik

Magnusson MS (1996) Hidden real-time patterns in intra- and inter-individual behavior: Description and detection. Eur J Psychol Assess 12(2):112–123

Magnusson MS (2000) Discovering hidden time patterns in behavior: T-patterns and their detection. Behav Res Methods, Instrum, Comput 32(1):93–110

Magnusson MS (2005) Understanding social interaction: Discovering hidden structure with model and algorithms. In: Anolli L, Duncan S, Magnusson M, Riva G (eds) The hidden structure of interaction: From genomics to culture patterns. IOS Press, Amsterdam, pp 4–24

Magnusson MS (2006) Structure and communication in interaction. In: Riva G, Anguera M, Wiederhold B, Mantovani F (eds) From Communication to presence: Cognition, emotions and culture towards the ultimate communication experience. IOS Press, Amsterdam, pp 127–146

Magnusson MS (2016) Time and self-similar structure in behavior and interactions: From sequences to symmetry and fractals. In: Magnusson M, Burgoon J, Casarrubea M (eds) Discovering hiden temporal patterns in behavior and interaction. Springer, New York, pp 3–35

McNeill D (1981) Action, thought, and language. Cognition 10:201–208

Mondada L (2007) Multimodal resources for turn-taking: Pointing and the emergence of possible next speakers. Discourse Stud 9(2):194–225

Mondada L (2013) Conversation analysis: Talk and bodily resources for the organization of social interaction. In: Müller C, Cienki A, Fricke E, Ladewig S, McNeill D, Teßendorf S (eds) Body–Language–Communication. An International Handbook on Multimodality in Human Interaction, vol. 1, De Gruyter, Frankfurt, pp 218–227

Müller C (2003) O the gestural creation of narrative structure: A case study of a story told in a conversation. In: Rector M, Poggi I, Trigo N (eds) Gestures, meaning, and use. Universidade Fernando Pessoa Press, Porto, pp 259–265

Müller C (2009) Gesture and language. In: Malmkjaer K (ed) The Roudledge linguistics enciclopedia. Routledge, London, pp 510–518

Müller C (2013) Gestures as a medium of expression: The linguistic potential of gestures. In: Müller C, Cienki A, Fricke E, Ladewig S, McNeill D, Teßendorf S (eds) Body–language–communication. An international handbook on multimodality in human interaction, vol. 1 ., De Gruyter, Frankfurt, pp 202–217

Norris S (2013) Multimodal (inter)action analysis: An integrative methodology. In: Müller C, Cienki A, Fricke E, Ladewig S, McNeill D, Teßendorf S (eds) Body–language–communication. An international handbook on multimodality in human interaction, vol. 1, De Gruyter, Frankfurt, pp 275–286

Ochs E (1979) Transcription as theory. In: Ochs E, Schieffelin B (eds) Developmental pragmatics. Academic Press, New York, pp 43–72

O'Connell DC, Kowal S (2000) Are transcripts reproducible? Pragmatics 10:247–269

Portell M, Anguera MT, Hernández-Mendo A, Jonsson GK (2015) Quantifying biopsychosocial aspects in everyday contexts: an integrative methodological approach from the behavioral sciences. Psychol Res Behav Manag 8:153–160

Poyatos F (1986) Enfoque integrativo de los components verbales y no verbales de la interacción y sus procesos y problemas de codificación. Anuario de Psicología 34(1):125–155

Poyatos F (2002a) Nonverbal communication across disciplines, volume I: Culture, sensory interaction, speech, conversation. John Benjamins, Amsterdam

Poyatos F (2002b) Nonverbal communication across disciplines, volume II: Paralanguage, kinesics, silence, personal and environmental interaction. John Benjamins, Amsterdam

Reynar JC (1998) Topic segmentation: Algorithms and applications. Unpublished doctoral dissertation. institute for research in cognitive science, University of Pennsylvania. http://repository.upenn.edu/ircs_reports

Sacks H (1972) An initial investigation of the usability of conversational data for doing sociology. In: Sudnow D (ed) Studies in social interaction. Free Press, New York, pp 31–74

Sacks H, Schegloff E, Jefferson G (1974) A simplest systematics for the organization of turn-taking for conversation. Language 4:696–735

Sánchez-Algarra P, Anguera MT (2013) Qualitative/quantitative integration in the inductive observational study of interactive behaviour: Impact of recording and coding predominating perspectives. Qual Quant Int J Methodol 47(2):1237–1257

Scheflen A (1966) Natural history method in psychotherapy: Communicational research. In: Gottschalk L, Ayerbach A (eds) Methods in research in psychotherapy. Appleton-Century-Crofts, New York, pp 263–289

Schegloff EA (1984) On some gestures' relation to talk. In: Atkinson J, Heritage J (eds) Structures of social action. Cambridge University Press, Cambridge, pp 266–296

Schegloff EA (2000) On granularity. Ann Rev Sociol 26:715–720

Schegloff EA (2005) On integrity in inquiry...of the investigated, not the investigator. Discourse Stud 7(4–5):455–480

Schegloff EA, Sacks H (1973) Opening up closings. Semiotica 8:289–327

Selting M (2013) Verbal, vocal and visual practices in conversational interaction. In: Müller C, Cienki A, Fricke E, Ladewig S, McNeill D, Teßendorf S (eds) Body–language–communication. An international handbook on multimodality in human interaction, vol. 1, De Gruyter, Frankfurt, pp 589–609

Sidnell J (2006) Coordinating gesture, talk, and gaze in reenactments. Res Lang Soc Interact 39(4):377–409

Sidnell J (2013) Basic conversation analytic methods. In: Sidnell J, Stivers T (eds) The handbook of conversation analysis. Wiley-Blackwell, New York, pp 77–99

Sidnell J, Stivers T (eds) (2013) The handbook of conversation analysis. Wiley-Blackwell, New York

Sperber D, Wilson D (1986) Relevance: Communication and cognition. Harvard University Press, Cambridge, MA

Stevens SS (1946) On the theory of scales of measurement. Science 103:677–680

Suárez N, Sánchez-López CR, Jiménez JE, Anguera MT (2018) Is reading instruction evidence-based? analyzing teaching practices using t-patterns. Front Psychol 9:7. https://doi.org/10.3389/fpsyg.2018.00007

Vaimberg R (2010) Psicoterapias tecnológicamente mediadas [Technology-mediated psychotherapy]. Unpublised Doctoral Dissertation. Universidad de Barcelona, Barcelona, Spain

Weick KE (1968) Systematic observational methods. In: Lindzey G, Aronson E (eds) Handbook of social psychology, vol II. Reading, Addison-Wesley, Mass., pp 357–451

White CH, Burgoon JK (2001) Adaptation and communicative design: Patterns of interaction in truthful and deceptive conversation. Hum Commun Res 27:9–37

Chapter 4
Research Methods for Studying Daily Life: Experience Sampling and a Multilevel Approach to Study Time and Mood at Work

Mariona Portell, Robin M. Hogarth and Anna Cuxart

Abstract The Experience Sampling Method (ESM) allows the examination of on-going thoughts, feelings and actions as they occur in the course of everyday life. A prime benefit is that it captures events in their natural context, thereby complementing information obtained by more traditional techniques. We used ESM to study time and mood at work. Our data were collected by sending 30 text messages over 10 working days to each of 168 part-time workers. On each occasion, respondents assessed their mood. We explored the joint effects of three sets of variables: activities in which people are engaged; individual differences; and time (i.e., when mood is measured). Since the data in our study can be thought of as being collected at two levels, we applied techniques of hierarchical linear models. The results indicated that activities were significant but no systematic individual differences were detected. There were some small diurnal effects as well as an overall "Friday effect." Lastly, the weather had little or no influence on self-reported mood state. We discuss the results in terms of their methodological implications for studying daily life.

4.1 Introduction

Research methods for studying daily life (hereafter RMSDaLi) encompass studies that use systematic, repeated measurements of biopsychosocial variables collected in situ (Mehl and Conner 2012; Portell et al. 2015c). The RMSDali are extending the research agenda in different applied fields (see, e.g., Bakker and Daniels 2013; Bell et al. 2017; Cordier et al. 2014; Os et al. 2017; Zembylas and Schutz 2016), and

M. Portell (✉)
Universitat Autònoma de Barcelona, Campus de Bellaterra, 08193 Cerdanyola del Vallés, Spain
e-mail: mariona.portell@uab.cat

R. M. Hogarth
Department of Economics and Business, Universitat Pompeu Fabra, Ramon Trias Fargas 25-27, 08005 Barcelona, Spain

A. Cuxart
Universitat Pompeu Fabra (retired professor), Ramon Trias Fargas 25-27, 08005 Barcelona, Spain

© Springer Nature Switzerland AG 2020
L. Hunyadi and I. Szekrényes (eds.), *The Temporal Structure of Multimodal Communication*, Intelligent Systems Reference Library 164,
https://doi.org/10.1007/978-3-030-22895-8_4

their basic requirements are closely linked to the context and times at which the data are collected.

The first requirement of RMSDaLi is for the data to be collected in a natural, real-world context. This ensures ecological validity and permits the study of inter-relations between experiences, everyday behaviors, and the setting in which these occur (Santangelo et al. 2014). The second is that events must be recorded as soon as possible after they occur. This is particularly important when self-report methods are used, as the longer the interval between the event and its registration, the greater the risk of retrospective bias (Trull and Ebner-Priemer 2014). A third requirement is that information should be recorded over a period of time (Shiffman et al. 2008; Moskowitz et al. 2009). An intensive recording across time offers a wealth of op-portunities to study within individual processes, and to study microprocesses (e.g., interrelationships between cognitive, affective, behavioral, and physiological vari-ables for short intervals of time, using T-pattern analysis; Magnusson et al. 2016). Evidently, an aspect that influences the validity of RMSDaLi is how the data collec-tion times are selected (Fahrenberg et al. 2007). The minimum requirement is that data should be collected systematically, taking into account: (1) the study objectives; (2) the distribution of the target behaviors over time; (3) susceptibility to retrospec-tion bias; (4) participant burden (e.g., time needed, extent to which the protocol interferes with intimacy); and (5) expected adherence to protocol (e.g., activation of recording devices).

One of the most influential RMSDaLi on the psychology of emotion is the Ex-perience Sampling Method (ESM) (Hektner et al. 2007), a method whose historical roots lie in the principles of representative design advocated by Brunswik (1944, 1956), see also Hogarth (2005), Portell et al. (2015b).

The aim of this chapter is to present an adaptation of ESM for studying time and mood at work with a data collection procedure that requires minimum technological resources. In addition, we combine this data collection procedure with a data analytic strategy based on multilevel modeling.

This multilevel modeling (MLM, also called the hierarchically linear model and random coefficient model) is one of the most recommended analytic approaches for ESM data (Conner and Mehl 2015; Mehl and Conner 2012; Santangelo et al. 2014). This is because MLM analyzes all people's data simultaneously to test for within—and between-person patterns, and it can (Conner et al. 2009): (1) model a relationship within each person's set of data points; (2) test whether those within-person patterns are the same or different across people; and, (3) if different, test whether other between-person variables (e.g., demographics, weather or personality) might account for the variance.

This chapter is organized as follows. We start by introducing the challenge of un-derstanding variability in levels of mood and happiness in daily life. Then we review the literature on the effects of time in studies of mood that provide the motivation for our analyses. This is followed by a description of our study, focusing on the participants and procedures used for data collection. Our results are presented in the subsequent section. Next, we outline the steps taken to establish the validity and reli-ability of our single mood measure. In the final sections, we discuss the implications of our findings and conclude by advocating methodological complementarity in the study of biopsychosocial variables in everyday contexts.

4.2   The Challenge to Study Time and Mood at Work

Understanding variability in levels of mood and happiness in daily life is an important topic that has attracted a significant scientific literature (see, e.g., Bradburn 1969; Csikszentmihalyi 1990; Strack et al. 1991; Diener and Seligman 2004; Kahneman et al. 2004). This variability can be viewed as being moderated by three classes of variables. First are people's activities and the occurrence of specific events (see, e.g., Csikszentmihalyi 1990; Clark and Watson 1988; Kahneman et al. 2004). Second are individual differences such as age, gender, culture, and personality (see, e.g., Diener et al. 2003; Oishi et al. 2007). And third are time-related factors that lie outside individual control and which form the background against which daily life is lived.

Our purpose here is to explore—*within the same investigation*—the joint effects on mood at work of three time-related variables: time-of-day, day-of-the-workweek, and weather. For example, Murray et al. (2009) document how the human circadian system modulates positive mood across the day. In short, every day involves a recurring cycle of 24 h involving day (light) and night (dark) and humans typically sleep in the latter period with consequent effects on energy levels. Days too are organized in cycles of seven (perhaps artificially), and weather follows seasonal patterns with random variations within seasons.

Two main issues motivate our study. The first concerns the importance of the timing of measures of social well-being (or happiness). Does it matter *when* such judgments are elicited? While it is well-known that mood assessment can be affected by, say, question order (see, e.g., Strack et al. 1988) or the occurrence of major events (positive or negative), the role of temporal variables is not clear. Moreover, if effects exist, how important are they and should investigators be concerned about "bias" in responses? The second issue is to illuminate the joint effects of different temporal variables on mood. Are there regularities? Are some more important than others?

Our data were collected in two studies that used the Experience Sampling Method (ESM) to investigate everyday perceptions of risk (Hogarth et al. 2007; Hogarth et al. 2011). A key feature of both studies was that the first question respondents were asked when prompted at random moments was an assessment of mood. Actually, the first three questions of both studies were identical across experimental treatments (the second and third questions asked what the participants were doing and whether the activity was personal or professional). Thus, since in the analyses reported here we only use the first three responses, we aggregate the two sets of data.

Respondents completed prepared response sheets when triggered by text messages sent to their cellular telephones at *random* moments during their working days. Given our research objectives, this methodology had two major advantages. First, by having respondents answer questions when prompted by text messages, we controlled the timing of our measurements thereby avoiding possible biases of systematic data collection methods such as diaries. Second, there were two consequences of sending requests for responses on a random basis. One was that we collected random samples of each respondent's mood states. The other was the lack of systematic bias in the times when mood state was elicited.

We also gathered data on what respondents were actually doing when asked to answer questions. Thus, a further innovative feature of our data collection and analysis is the joint consideration of effects on mood due to the three classes of variables discussed above, namely activities, individual differences, and temporal factors.

Unlike much of the literature on mood, we used a single bipolar measure. We simply asked respondents "*How would you evaluate your emotional state right now?*" on a scale from 1 (very negative) to 10 (very positive). While this "overall mood" question does not distinguish between negative and positive moods (Watson and Tellegen 1985; Clark and Watson 1988) nor different types of moods (see, e.g., Stone et al. 1996), it does provide a simple overall measure to which our respondents could relate easily in the context of the other questions they were asked. In addition, we note that the use of single questions of "subjective well-being" is common in many happiness surveys and has provided meaningful data (see, e.g., Frey and Stutzer 2002; Diener and Seligman 2004). As such, answers to our question can be thought of as summary measures of overall mood, possibly equivalent to a ratio of positive to negative moods. Later in this chapter, we detail steps we took to assess the validity and reliability of our single mood measure.

4.3   The Role of Temporal Factors

Earlier studies investigated effects of time on mood. Of particular importance are those due to the time-of-day (diurnal), day-of-the-week, weather, and season. However, since our data does not contain sufficient samples of seasonal observations, we shall exclude the latter.[1]

4.3.1   Diurnal Effects

The existence of cyclical patterns for several types of mood has attracted the attention of researchers using a wide variety of methods (Kahneman et al. 2004; Murray et al. 2009). A priori, this is not a simple area of investigation in that natural biological cycles can be masked by factors such as the social organization of the day and specific events (cf., Clark and Watson 1988). Thus, in an especially interesting study where a heterogeneous sample of 18 adults were kept in isolation over five days, Monk et al. (1985) measured several moods and activities at frequent intervals. Their measures

---

[1] Seasonal effects of weather on moods and behavior have been documented (see, e.g., Smith 1979; Harmatz 2000).

of "happy" (or positive mood) and overall "wellbeing" showed inverted-U patterns with the maxima being achieved some 4.1 h after waking. "Sad" (or negative mood) had no temporal pattern.

Wood and Magnello (1992) had several different groups of respondents (students and non-students) assess moods and energy levels at different times of the day. Their conclusions were, in brief, that positive mood is affected by the diurnal cycle but negative mood did not. Second, moods with cycles reached their peaks between 10 a.m. and noon, and although energy levels dipped after lunch, they rose late at night for students. Third, they speculated that whereas positive moods might have a biological component, negative moods might reflect environmental factors to a greater extent. In a related study of chronic fatigue syndrome patients and a control group, Wood et al. (1992) again found that diurnal patterns of energy were highly correlated with positive mood and reached their peaks between 10 a.m. and noon but measures of negative affect had no diurnal pattern.

Further evidence for the inverted-U shaped curve across the day for positive affect—and no relation for negative affect—can be found in several other studies (Thayer 1987; Clark et al. 1988; Watson et al. 1999; Murray et al. 2002, 2009; Peeters et al. 2006).

Stone et al. (1996) made a detailed study of the moods experienced by 94 employees of a large insurance company in New York. They collected data every 15 min over the course of one working day using a diary method. They found that moods were influenced by specific activities or location that were correlated with times in the day (such as commuting in early morning/late afternoon or lunch at noon), but that nonetheless other diurnal cycles were not dependent on such factors (in particular, "rushed," "sad," and "tired").

Stone et al. (2006) analyzed a large dataset involving responses by 909 working women in Texas (Kahneman et al. 2004). They tabulated changes in twelve moods (assessed by adjectives) across one working day and noted several distinctive diurnal patterns. There were peaks for positive emotions at noon and in the evening and peaks for negative emotions in mid-morning and mid-afternoon. Other moods had V and inverted-U shaped patterns ("tired" and "competent," respectively). The advantage of the diary methodology used by Stone et al. (2006) was that it captured much relevant data. However, these were limited to the activities of a single day and thus could not capture variation in factors such as the weather. Nevertheless, as the authors themselves state:

> With regard to the diurnal cycles observed in this sample of Texas women, not only were several findings based on smaller scale studies replicated, we detected diurnal rhythms that to our knowledge have not previously been reported. A consistent and strong bimodal pattern was found for positive and negative emotions. For the three positive adjectives, emotion levels during the work day had a peak at noon and a second peak starting at about 7 p.m. and the higher level lasted the rest of the evening. Conversely, peaks for the six negative adjectives were at about 10 a.m. and then at 4 or 5 p.m., although this pattern was relatively weak for some of the adjectives. One interpretation of this bipolarity is that the elevation of negative emotions was due to work and that lunchtime provided a respite from the demands of the work environment, reducing negative emotions (and increasing positive emotions)...(Stone et al. 2006, p. 145).

4.3.2   Day-of-the-Week Effects

Most people are familiar with feelings of "blue Mondays" and "happy Fridays" (TGIF) as markers of starting and ending the work week. However, what evidence exists to support these notions?

Rossi and Rossi (1977) reported a study of daily moods of university students over a 40-day period. Using a measure of the ratio of the endorsements of positive to negative mood adjectives, they found an increasing trend in mood from Monday through Friday with a stronger slope for men ($n = 15$) than women ($n = 67$). They explain this gender effect by noting that women's daily moods are confounded by effects of menstrual cycles that do not match days of the week. However, they also show that there are day-of-the-week effects for women that influence the effects of menstrual cycles.

In another study involving undergraduate students (39 females and 35 males) who completed mood reports for 84 consecutive days, Larsen and Kasimatis (1990) found a strong weekly pattern of data similar to that of Rossi and Rossi (1977). Moreover, they detected a systematic personality difference in that extraverts exhibited more variability in daily moods than introverts.

Replication of these effects with larger and more representative samples has, however, not proven successful. For example, Stone et al. (1985) carried out several studies with quite big samples of married men. Their findings can be summarized by stating that although their respondents believed that Mondays were "blue" and Fridays "happy," this was not the case when mood was actually measured on those days. (At weekends, however, positive mood was generally higher and negative mood lower.) In a diary study involving 166 married couples over six weeks, Bolger et al. (1989) found no day-of-the-week effects. However, from their study one might also infer that these could have been perturbed by other, more significant events.

4.3.3   Weather Conditions

Most people have an intuitive feeling that mood levels vary with weather. However, both mood and weather conditions can be classified in terms of different parameters, and the empirical research does not offer a clear picture.

Several studies clearly show effects of weather on human actions where it is assumed that mood, as a reaction to changes in weather, affects behavior. For example, Hirshleifer and Shumway (2003) showed that there is a significant, positive correlation between the amount of sunshine and stock returns. Moreover they documented this across 26 countries (national exchanges) from 1982 to 1997, thereby providing support for an earlier study by Saunders (1993) in the US (see also Trombley 1997). Further evidence has been provided by Rind (1966) and Rind and Strohmetz (2001) who documented how beliefs concerning good weather increased tips given in restaurants. Lastly, Simonsohn (2007) reported that weather affects the decisions of university admissions officers. Academic attributes are weighted more heavily when there is cloud cover (i.e., lack of direct sunshine).

There is some evidence that sunshine has a direct effect on mood (broadly defined). High levels of sunshine have been seen to increase self-reports of happiness (Schwartz and Clore 1983) and similar effects on mood have been reported by Cunningham (1979) and Parrott and Sabini (1990). Despite this, Schkade and Kahneman (1998) found no differences in the life satisfaction of students in two regions in the US that differed in desirable weather (the Midwest and Southern California). However, when respondents rated life satisfaction of a similar type in the other region, Midwesterners gave higher ratings to Californians than they did themselves.

In an early study involving relatively few observations, Goldstein (1972) reported that better mood was associated with high barometric pressure on some measures but low on others. In addition, his results suggested that gender and being an external (on Rotter 1966 IE scale) might mediate reactions between mood and weather. A decade later and using a larger sample, Sanders and Brizzolara (1982) concluded that the effect of weather on mood is most marked by levels of humidity (better moods being associated with low humidity). This result was replicated by Howarth and Hoffman (1984) who related measures of ten mood variables to eight weather variables collected from 24 male respondents over eleven days. Humidity, temperature, and hours of sunshine were found to have the greatest effects on mood. However, humidity was the most significant "predictor" (in a regression and canonical correlation analysis).

Denissen et al. (2008) conducted a comprehensive online diary study ($N = 1,233$) that examined possible effects of six weather parameters (temperature, wind strength, hours of sunshine, precipitation, air pressure, and hours of daylight) on three measures of mood (positive affect, negative affect, and tiredness). Using multilevel approach, they found no significant effects of the daily weather on positive affect. There were negative effects of temperature, wind power and sunshine, and sunshine also resulted in tiredness. However, overall weather fluctuations accounted for little variance in mood. Interestingly, Denissen et al. (2008) reported individual effects, but these could not be explained by either personality (the Five Factor model) or gender.

In a study by Keller et al. (2005), no correlation was found between weather and mood at different times of the year except that pleasant weather (high temperature or barometric pressure) was related to a better mood during the spring as the time spent outdoors increased. These investigators posit a post-winter contrast effect due to the time spent outdoors in more pleasant conditions.

At one level, it is surprising that the literature does not provide evidence of "simpler" effects of weather on mood. However, both weather and mood are multidimensional and, in addition, the studies reviewed employed a variety of different methodologies. Also, the sampling of weather occurred at different times of the year and in different geographical locations. Lastly, people who have experienced different weather conditions throughout their lives might well react in different ways.

4.3.4   Other Factors

After assessing their own mood states, our respondents were asked what they had just been doing (see below). Thus, we can also investigate to what extent current activities can affect mood. Three types of variables are of interest: (1) the kind of tasks participants were performing (recall they were students and part-time workers questioned mainly while at work); (2) whether they were doing something that was personal or professional. The literature suggests that people involved in "desirable" events exhibit better moods than those who are not so involved (David et al. 1997); and (3) whether they were doing something on their own or with others (the latter has been shown to be associated with better moods, Clark and Watson (1988)).

4.4   The Study

The data were collected in two phases. The first took place in February and May of 2005, the second in October of 2006. Each phase involved a separate ESM study designed to illuminate the perception of risks (Hogarth et al. 2007, 2011) but, as noted above, since the first three questions were identical in both, we combined the two datasets for the purpose of the present analysis (which only involves these three questions).

4.4.1   Participants

All participants were students recruited from the Universitat Autnoma de Barcelona. A condition of their participation was that they had part-time jobs (defined by at least one third of full working days). There were 168 participants in all—74 in phase 1 and 94 in phase 2. There were more women than men—46 versus 28 in phase 1, and 64 versus 30 in phase 2. They ranged in age between 17 and 56 with a median of 22 in phase 1, and 19 in phase 2. Those participating in phase 1 were each paid 30 euros. In phase 2, the remuneration was 35 euros. Participants were required to respond to the questions detailed below as well as to some other questions that are not relevant to this analysis. In addition, they were required to attend sessions before and after the study for instructions and debriefing (which included some further questions).

4.4.2   Procedure

We sent text messages to participants between 8 am and 10 pm over a two-week period that excluded week-ends, i.e., for 10 consecutive working days. The objective was to send participants messages during the part of the day in which they were mostly at work. Thus, depending on their working hours, some participants received their messages between 8 am and 3 pm and the others between 3 pm and 10 pm (43 and 125 participants, respectively). To determine when messages should be sent, we divided time into segments of 15 min and chose six segments at random each day (three for each group of participants). When they received a message, participants were required to note the date and time and to answer a series of questions (all questions were asked in Spanish). The first three questions and types of scale used were:

1. *How would you evaluate your emotional state right now?* Scale from 1 (very negative) to 10 (very positive).
2. *What are you doing right now?* Open-ended and subsequently referred to as ACT (ACTivity).
3. *Is ACT professional or personal in nature?* Binary response, coded (0/1).

There were up to five additional questions afterwards that varied according to phase and experimental conditions within phases (Hogarth et al. 2007, 2011).

After completing the task, participants were thanked, debriefed, and paid in a post-experimental session in which they also answered demographic and other questions. Phase 1 participants also completed (Rotter 1966) Internal-External "Locus of Control" questionnaire (IE).

4.4.3 Data Analysis

The design of our study involved data that could be viewed as being collected at two levels. One of these levels—called level 1—is represented by participants' responses to the 30 occasions on which they received text messages (i.e., at the level of occasions). The other—level 2—is that of the participants themselves (i.e., characteristics of the participants that do not change across the 30 occasions). Thus, for example, it is of interest to know whether, say, mood at the moment judgments are elicited (question 1) is associated with what the participants were doing (question 2)—i.e., at level 1—and also whether such judgments reflect differences between the participants in, say, gender—i.e., at level 2. As such, our data can be efficiently modeled using the techniques of hierarchical linear models (HLM) (Byrk and Raudenbush 2002; Goldstein 1995; Longford 1993).

4.5   Results

4.5.1   Response Rates

Out of the 5,040 (= 168 × 10 × 3) messages sent, 5,022 were received (99.6%). For various reasons, people might not receive text messages when they are sent (e.g., cell telephones may have been turned off). We therefore checked the extent to which messages were received when they were sent. In phase 1, participants reported receiving messages between zero and 22 min after they were sent with an overall mean (median) of 3 (2) minutes. In phase 2, the range lay between zero and ten minutes after reception (with an overall mean of one minute). We deem both response rates and reported times of receiving messages satisfactory.

4.5.2   An Overview

Table 4.1 lists the results of the analysis of a hierarchical model and provides an overview of our findings. Here, we present six models to illustrate the effects of different classes of variables.

Model 1 just estimates the overall mean of mood (testing for possible effects due to phase) without accounting for any other factors and the residual variance at level 2 and at level 1 (between and within individuals, respectively). The estimate for the overall mood is 6.76 on a scale from 1 ("very negative") to 10 ("very positive"). The intraclass correlation is 0.20, meaning that 20% of the total variance in mood is accounted for by individual differences. Model 2 shows a statistically significant and fairly large increase in mood (0.46 points) due to personal, rather than professional activities. In Model 3, significant effects of different types of activities, and the extent to which they involve interaction with other people, are estimated. Model 4 shows diurnal effects. Model 5 includes the latter due to the days of the week, and Model 6 captures the effects due to weather.

It should be added that Table 4.1 provides an overview of all of our data and that all the models have been estimated assuming fixed effects. We have also estimated models that assumed random effects and, in our discussion of results for each class of variables below, we describe implications of different ways of analyzing the data.

4.5.3   Level-2/Personal Variables

There are two kinds of level-2 variables, namely, methodological and personal. For the former, we recall that the study was conducted in two phases and, within phases, participants answered questions mainly in the morning or in the afternoon. As shown in Table 4.1, the dummy variable for phase 2 is not significant and justifies aggregating

Table 4.1 The role of activities and time in mood assessment

	Model 1		Model 2		Model 3		Model 4		Model 5		Model 6	
	Initial model		Type of activity		Different activities		Time of day		Day of workweek		Weather	
Fixed effects	Coefficient	t-ratio	Coefficient	t-ratio	Coefficient	t-ratio	Coefficient	t-ratio	Coefficient	t-ratio	Coefficient	t-ratio
Level 2 variables												
Intercept	**6.76**	68.97	**6.48**	62.74	**6.47**	62.22	**6.19**	41.80	**6.13**	39.36	**6.10**	35.45
Phase 2	−0.11	−0.84	−0.16	−1.18	−0.21	−1.62	−0.23	−1.68	−0.23	−1.63	−0.12	−0.80
Level 1 variables												
Personal (personal = 1, professional = 0)			**0.46**	9.10	**0.33**	5.61	**0.31**	5.18	**0.32**	5.33	**0.32**	5.16
Eating and drinking					**0.34**	4.41	**0.33**	4.32	**0.32**	4.15	**0.34**	4.26
Entertainment					**0.31**	3.72	**0.30**	3.61	**0.31**	3.70	**0.30**	3.54
Personal care/rest/sleep					−0.17	−1.97	−0.12	−1.32	−0.13	−1.49	−0.11	−1.20
Interacting with												
Family/friends					**0.81**	9.15	**0.79**	9.00	**0.76**	8.65	**0.76**	8.53
Children (professional)					0.36	2.57	0.35	2.45	0.36	2.53	0.34	2.35
Time of the day (ref. 8:00–10:20)												
10:21–12:40							0.30	2.75	0.31	2.80	0.26	2.32
12:41–15:00							**0.40**	3.70	**0.44**	3.99	0.35	3.02

(continued)

Table 4.1 (continued)

	Model 1	Model 2	Model 3	Model 4		Model 5		Model 6	
	Initial model	Type of activity	Different activities	Time of day		Day of workweek		Weather	
15:01–17:20				0.25	1.61	0.28	1.75	0.08	0.50
17:21–19:40				0.30	1.87	0.29	1.78	0.09	0.53
19:41 and after				0.38	2.39	0.36	2.28	0.18	1.07
Day of the workweek (ref. Monday)									
Tuesday						−0.00	−0.02	−0.02	−0.34
Wednesday						−0.08	−1.13	−0.08	−1.07
Thursday						0.09	1.30	0.04	0.51
Friday						**0.28**	3.95	**0.25**	3.42
Weather (Sunshine hours)								0.02	2.13
Random effects									
Level 2 (individuals)									
Residual variance	**0.633**	**0.641**	**0.646**	**0.649**		**0.649**		**0.670**	
Level 1 (occasions)									
Residual variance	**2.458**	**2.417**	**2.351**	**2.344**		**2.330**		**2.344**	

Note Coefficients/variance components significant at $p<0.001$ are **bold**, significant at $p<0.05$ are underlined using t-tests or chi^2 as appropriate the data from both phases. Nor is there a main effect for responding in the morning or the afternoon but this distinction does interact with level 1 variables (see *Activities* below).

No effects for gender are shown in Table 4.1 because there were none. As to possible effects of personality, we did have measures of Rotter (1966) IE scale ("Locus of Control") for the 74 participants in phase 1. Interestingly, while the correlation between IE scores and mean mood was not significant ($r = -0.16$), the correlation between IE scores and the standard deviation of mood was ($r = 0.34$, $p = 0.003$), and for women in particular ($r = 0.42$, $p = 0.004$). Variability in mood, then, seems to be associated with more externally-oriented personalities (especially for women). However, it is not clear how this squares with previous studies on locus of control (see, e.g., Blair et al. 1999; Klonowicz 2001).

4.5.4 Activities

In our study, respondents reported on activities in their own words in response to the second question they received (i.e., after reporting mood). We classified these data as follows. First, for data from phase 1 we established definitions of categories for the activities. Then, two researchers independently allocated responses to categories ($Kappa = 0.65$). Disagreements between the two coders were resolved by having them discuss until they reached consensus. Second, for data from phase 2, two coders were trained in the use of the categories employed in phase 1. They then independently allocated responses to categories and discussed disagreements with a third person (overall $Kappa = 0.95$). As a third step, all the data for professional activities (phases 1 and 2) were submitted to an additional analysis to determine more specific categories.

Consistent with the literature, our data show variation in mood by the activities in which respondents were engaged (cf., Kahneman et al. 2004). First, as shown by Model 2, being involved in personal as opposed to professional activities has a positive impact (cf., David et al. 1997). Furthermore, several activities have significant coefficients in Model 3 of Table 4.1—in particular "Eating and drinking," "Entertainment," and "Personal care/rest/sleep"—that are over and above the effect of the dummy variable for "personal/professional."[2] In addition, there is a strong effect (0.81) for interacting with family/friends (see also Clark and Watson 1988). Further insight is provided by Fig. 4.1 that shows 95% confidence intervals of mean z-scores for mood broken down into the categories of activities that we established and it highlights the distinction between personal and professional types of activity.[3] These show that, relative to each respondent's average mood state, professional ac-

---

[2]The reference category used as a base for coding the dummy variables for different activities was "Housework, personal time organization, and managing funds."

[3]We calculated z-scores for each individual respondent such that the mean of each person's mood judgments was 0 with a standard deviation of 1. This allowed us to categorize all obser-

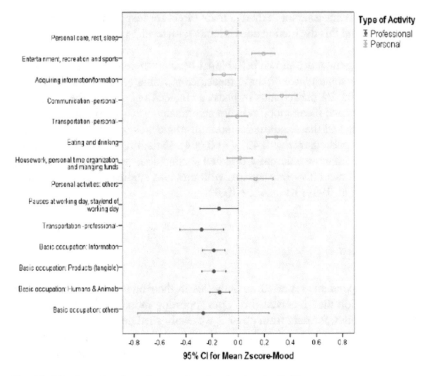

Fig. 4.1 Mood as a function of personal and professional activities

tivities were generally associated with a negative (i.e., below average) mood whereas most personal activities were generally above average.

A more detailed analysis of the Table 4.1 data broken down by whether participants were working mornings or afternoons reveals an interaction with the type of activity (not shown in Table 4.1). Specifically, whereas the pattern of significant effects for different activities for afternoon workers is the same as that for the whole sample, this is not of the case for morning workers. For the latter, the coefficients for "Eating and Drinking" and "Entertainment" are not statistically significant nor are the coefficients for interacting with "family/friends" and "children". Still, the coefficient for "Personal care/rest/sleep" is significant (-0.64, $p < 0.001$).[4] A plausible interpretation is that the morning and afternoon groups differed in the nature of their activities.

---

vations/occasions as being positive or negative, i.e., whether they were above or below each individual's mean mood score.

[4] These results are robust to analyses assuming fixed or random coefficients.

4.5.5   Diurnal Effects

Model 4 of Table 4.1 suggests there are effects of the time of day on mood between 8 and 10:20 am during the working week. As can be seen, there are significant effects above this base level between 10:21 to 12:40 and 12.41 to 15:00. Then, there is a significant effect at the end of the day, i.e., from 19:41 and after. This pattern is illustrated in Fig. 4.2 which has 95% confidence intervals for mood at different times of the day. As mentioned, mood starts low in the morning, rises to the period between 12:41 to 15:00, and then falls off sharply in the afternoon before rising again in the evening.

Although this figure shows variations across the day, it is important to recall that the data are comprised of morning and afternoon groups such that the three earlier estimates are based predominantly on the morning group and the three later estimates on the afternoon group. Nonetheless, the pattern of data is remarkably similar to results reported by other researchers—for positive but not negative mood. (Reports of negative mood are that it is almost "flat" or "unpredictable" across the day.) Several studies mentioned above provide evidence of a similar inverted-U pattern prior to the evening when there is a late upturn (see, e.g., Monk et al. 1985; Wood and Magnello 1992; Wood et al. 1992; Peeters et al. 2006; Stone et al. 2006). A discrepancy with our data is that the mid-day peak appears later than in the other, mostly US, studies.

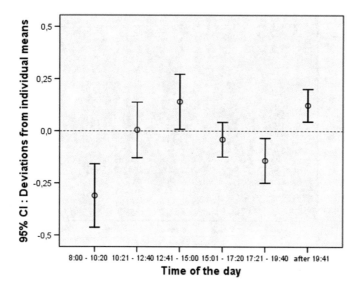

Fig. 4.2   Mood as function of time of day

There is a plausible cultural explanation. Whereas lunch usually starts at around 12 noon in the US, it is much later in Spain, starting at 2 or even 3 pm. This point then appears to displace the diurnal pattern.

In short, we find a diurnal pattern that is consistent with data from other studies involving positive mood as well as energy levels.

4.5.6 Day-of-the-Workweek Effects

Model 5 of Table 4.1 suggests there is a day-of-the-working week effect—with a higher mood on Fridays. This is also shown graphically in Fig. 4.3. As noted above, when one excludes weekends, effects for day-of-the-week are not a consistent finding. Our data do support the findings by Rossi and Rossi (1977) for a Friday effect in a part-time worker population. When comparing the morning and afternoon groups, the patterns of day-of-the-week effects are remarkably similar except that the Friday effect was only slightly greater for the morning group than for the afternoon group (not shown in Fig. 4.3).

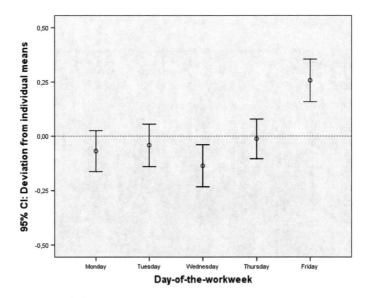

Fig. 4.3 Mood as a function of day-of-the-workweek

4.5.7 Weather Conditions

We examined meteorological conditions for the dates when our data were collected and we identified 10 different measures.[5] Of these, only one variable—daily sunshine (total number of hours)—was statistically significant as shown by Model 6 of Table 2.1 ($t = 2.13$, $p < 0.05$). However, when the same coefficient is estimated with robust standard errors, its significance can be questioned ($t = 1.56$, $p = 0.12$). More importantly, whether statistically significant or not, the effect is small.

Overall, our data do not show effects of variation in weather on mood. This only adds to the confusion on this topic in the literature. One explanation for the lack of effects in our data might be due to the nature of the generally pleasant Mediterranean climate in the Barcelona area. Although the data collection took place in different months (February, May, and October), the latter two months are typically characterized by pleasant weather, and February is usually mild. If data collection had also taken place in July and August, it is possible that the unpleasant feeling could have been associated with high humidity s factor as well, as reported in other studies (Sanders and Brizzolara 1982; Howarth and Hoffman 1984).

4.6 Using a Single Measure of Mood

The use of a single measure of mood requires some justification. Here, we present four arguments.

First, recall that we elicited self-reported mood in an ESM study where, to avoid reactivity,[6] we limited the number of questions (Hektner et al. 2007; Larsen and Fredrickson 1999).

Second, we can ask whether the results we obtained have face validity or, more precisely, what Hektner et al. (2007) refer to as *situational validity*. In other words, are participants' reports of mood coherent with other more "objective" findings and data in the study? The answer is undoubtedly "Yes." Consider, for example, the findings reported above about better moods being associated with personal as opposed to professional activities as well as the diurnal and day-of-the-week effects.

Our third argument is that we do, in fact, have some data for phase 2 that could be considered complementary to mood. Specifically, after answering the three questions defined above (see Procedure), participants in this second phase were also re-

---

[5]These included: daily average temperature (C); precipitation (liters per square meter); rain (dummy variable, 1:yes; 0:not); daily sunshine (total number of hours); relative daily sunshine (percentage out of expected total hours); degree of cloudy at 7am (scale from 0 to 8); degree of cloudy at 1pm (scale from 0 to 8); daily solar radiation (watts per square meter); daily average of relative humidity (%); and daily average of barometric pressure (in hectoPascals, hPa). The data were obtained from the Servei Meteorolgic de Catalunya, Xarxa d'Estacions Meteorolgiques Automtiques (XEMA) del Valls Occidental and Observatori Fabra (Barcelona).

[6]We use the term "reactivity" to mean a phenomenon that occurs when individuals alter their performance or behavior due to the awareness that they are being observed.

Table 4.2 Correlations of mood with SAM measures

A. Correlations between individuals (n = 94)

		Valence[a]	Arousal[b]	Dominance[c]	Mood[d]
SAMs:	Valence	1.00			
	Arousal	0.07	1.00		
	Dominance	−0.21	0.32	1.00	
Mood		**−0.66**	−0.03	**0.41**	1.00

B. Correlations within individuals (n≥2.779)

		Valence[a]	Arousal[b]	Dominance[c]	Mood[d]
SAMs:	Valence	1.00			
	Arousal	0.07	1.00		
	Dominance	**−0.37**	0.01	1.00	
Mood		**−0.57**	−0.11	**0.33**	1.00

Note figures in **bold** indicate p<0.001
[a]Scale "happy" (1) left to "unhappy" right (9)
[b]Scale "aroused" (1) left to "quiet" (9) right
[c]Scale "lack of control" (1) left to "dominating" (9) right
[d]Scale "very negative" (1) to "very positive" (10)

quired to report feelings of their emotional states using the method of self-assessment manikins (SAMs, Bradley and Lang 1994). The SAMs represent visually three basic dimensions of emotions in reactions to events or situations. These are (a) valence (or pleasure), (b) arousal, and (c) dominance. Each emotion is captured by five "cartoon" impressions, going from one extreme to the other. For example, valence is represented in the form of five different figures (mainly faces) going from happy smiling to unhappy. For each of the three emotions, participants simply checked the figure—or between adjacent figures—that corresponded most to their feelings. Conceptually, one would expect valence to have some connection with our measure of mood, given that it taps into an intuitive sense of happiness. However, no relation would be expected between mood and arousal although there might be a relation with dominance (a better mood being associated with more control).

Table 4.2 lists correlations between our mood measure and the SAMs—both between individuals (A) and within individuals (B). First note that there are appropriate and significant correlations between mood and valence (a happier valence being associated with a more positive mood). There is no significant relation between mood and arousal; and there is a positive relation between mood and dominance (the better the mood, the more the dominance). However, note that dominance and valence are also correlated here.

We know that mood at a particular point in time is not the same thing as emotional reactions to a situation. However, since these are simultaneous expressions of affective states, we would expect coherence among the different measures. Thus, the pattern of correlations in Table 4.2 supports the hypothesis that our mood measure has an appropriate level of reliability as well as having convergent and discriminant validity.

Next, our fourth point relates to a question the participants of phase 1 answered in their post-experimental session. This was to assess their "emotional state over the last two weeks" using the same 1 to 10 scale ("very negative" to "very positive") as in the main study. While people's memories of their past average mood states might be biased, significant correlations between the given average and estimates of actual experience should provide further evidence of the reliability of the mood scale. In fact, this correlation (i.e.,between estimates of average mood over the previous two weeks and participants' recalled estimates) is 0.69 ($n = 74$, $p < 0.001$).

It has to be noted that this result resonates with the literature on the so-called "peak-end" rule where it has been found that memory of the experience of sequential events is better modeled by averaging the "peak" (i.e., most extreme) and "end" (i.e., last) stimulus as opposed to the average of all stimuli experienced (Frederickson and Kahneman 1993). However, when we applied the corresponding peak-end rule for our data, the correlation was lower than for the mean (i.e., 0.35 versus 0.69). There are alternative explanations. One is that, for whatever reason, the peak-end rule result does not apply to our data (see also, Kemp et al. 2008). The second is that whereas taking the mean of 30 randomly selected moments of experience provides an unbiased estimate of the average mood state, otherwise, the peak-end rule estimation could be misleading. That is, there is no guarantee that the sequence of stimuli sampled actually includes the most extreme experience (mood state) or, indeed, the most recent mood state. Lastly, we take heart from analytical results of Cojuharenco and Ryvkin (2008) who showed that, under many conditions, peak-end and average experience are quite highly correlated.

4.7   Discussion

We used the ESM to investigate assessments of mood made during working hours by 168 part-time workers on a simple bipolar scale—from 1 (very negative) to 10 (very positive)—on 30 different occasions across a period of 10 working days. We considered three classes of explanatory variables, namely, types of activities; individual differences; and temporal factors. We exploited the methodological potential of the ESM-HLM hybrid to explore joint effects of these explanatory variables and demonstrated how this can be used to illuminate the online analysis of individual-environment interactions in natural contexts.

The participants also reported what they were doing when their mood was assessed and we used these self-report data to classify their activities. We found there was a strong effect when activities were considered personal as opposed to professional (on average almost one-half point higher on the mood scale). Moreover, if personal activities involved eating and drinking, entertainment, or interaction with friends and family, assessments were even higher. These effects are similar to other studies (e.g., Clark and Watson 1988; Kahneman et al. 2004). We also analyzed our data for possible effects in the types of part-time work performed by our participants but, with the exception of a small positive effect for professional childcare (i.e., babysitting), we found no such differences. A disadvantage of our methodology is that it is, of course, ill-suited to capturing effects of unusual events or rare activities.

With the exception of gender, our participants were homogeneous with respect to age and other demographic features typical of a part-time student population. As such, one would not expect to find many effects due to individual differences. Moreover, except for Rotter's (1966) IE ("locus of control") scores for 74 of the 168 participants, we had no measures of personality. Nonetheless, our analysis highlighted two points. First, there were no principal effects or even interactions involving gender. Second, while IE scores did not correlate with mood, they did correlate with variability in mood in that more externally oriented participants had larger standard deviations of mood scores.

We investigated three types of temporal variable, namely, diurnal, day-of-the-workweek, and weather. Moreover, an important advantage of our methodology was that we could estimate simultaneously the potential effects of all these from the same data. Our results are mostly consistent with findings in the literature.

First, the diurnal pattern of our data suggests an inverted-U shape from morning until the early evening followed by a rise in the late evening—see Fig. 4.2. Such patterns have also been observed in studies that have examined feelings of positive mood (e.g., Stone et al. 2006). And, since negative mood appears to be unrelated to time across the day, the argument can be made that total mood (as either the sum or ratio of positive and negative mood) should also follow the pattern that we observed. Next, we note that some studies have identified diurnal mood levels (and energy) that differ by age of participants, with older people starting high (in the morning) and ending low at night and younger people having the reverse pattern (see, e.g., Wood and Magnello 1992). The pattern of our young, part-time worker population clearly followed that of younger people.

Second, we identified a Friday effect—see Fig. 4.3. As pointed out above, this is both consistent (Rossi and Rossi 1977) and inconsistent (Stone et al. 1985) with previous findings of day-of-the-week effects.

Third, we found little or no effect of the weather. This is consistent with findings concerning positive affect in the extensive, recent study by Denissen et al. (2008). However, we are acutely aware that our sample of Mediterranean weather may not have provided sufficient variation. In particular, our literature review suggested two variables that might be especially relevant to mood changes, namely; hours of sunshine, and humidity. It is possible that weather-related changes of mood might interact with individual differences. This should be investigated in future research (cf., Goldstein 1972).

Lastly, whereas our analyses did identify some statistically reliable effects of time, these were not large in the sense that they would make much difference in a predictive model. Often such a statement might be considered the "death knell" of a scientific investigation. However, this is not the case here. As stated at the outset, it is important to establish both the existence *and* size of effects. For example, from the viewpoint of research on subjective well-being at work, it is essential to establish the boundary conditions under which assessments of happiness are and are not subject to systematic influences. Thus it is useful to know that the main effects of temporal variables are small. Whether their interactions with other variablesand especially whether individual differences are negligible at all is an open question for future research.

Our study is not without methodological shortcomings. First, although the situations faced by our participants were sampled randomly, the participants themselves were selected by convenience only and hence we cannot assume that they are representative of the population as a whole. Second, by asking our participants to record their emotional reactions we might have unwittingly influenced their reports. Emotions have been shown to have unconscious effects (Ruys and Stapel 2008) and our study could have overlooked or misinterpreted these. Third, contrary to other ESM studies (Dubad et al. 2017) we did not use mood-monitoring apps for smartphones. Our study demonstrates the power of a very simple technology based on mobile phone and small paper-and-pencil response pads (providing the corresponding evidence on the data quality; see Hogarth et al. (2007), Hogarth et al. (2011). Clearly, the mood-monitoring apps for smartphones are an essential resource in the collection of ESM-data and they have demonstrated their superiority, especially in studies with young people (Liao et al. 2016). However, our simple approach might still remain an option for ESM-data collection in workplace studies where a procedure using paper and pencil could be considered easier, more acceptable or less intrusive than the touchscreen of the smartphone (e.g. studies with older workers, open-ended responses).

4.8 Conclusions

In this chapter, we provided an illustration of the methodological potential of the ESM-HLM hybrid for studying the ongoing relationship between individuals' experiences and behaviors as they occur in the course of everyday life.

We placed this ESM-HLM hybrid within the more general framework of the RMSDaLi. The explosion of information and communication technologies in recent years is stimulating the methodological innovations for quantifying biopsychosocial variables in everyday contexts (e.g. Doherty et al. 2014; Sandstrom et al. 2017). Each innovation usually solves some old problem but often presents new challenges. For RMSDaLi some of the open challenges are related to:

1. the appropriate benchmarks for assessing the validity of emotion-monitoring apps (Dubad et al., 2017);
2. a way to increase the participants' engagement and compliance to ESM protocols (whether they are digitally savvy or not), and the managing of noncompliance (Wen et al. 2017);

3. the proposal of more efficient designs to reduce the demands on participants (Silvia et al. 2014). Advances in these three areas will improve the ESM-HLM hybrid outlined in this chapter.

The study presented in this chapter is based on self-observation and self-reporting. As each assessment method has its own pros and cons, we strongly advocate the use of methodological complementarity for studying biopsychosocial fenomena in everyday contexts (Portell et al. 2015a). For example, the data presented in the previous sections could be complemented with biological markers (e.g. salivary cortisol) to get a better understanding of employees' daily response patterns to occupational stressors (e.g., Volmer and Fritsche 2016). It could also be complemented with data, that allow us to relate (self-observed) emotional variables with (hetero-observed) behavioral variables collected by designs grounded in observational methodology (Anguera 2003; Portell et al. 2015a). Another example of complementarity might be the mixing of the previous data with data obtained from qualitative interviews (e.g. to union representatives) following the mixed method approach (Creswell and Plano Clark 2011; Anguera et al. 2017). Using these approaches obviously increases the complexity of the research, but we do hope that future studies can strengthen the validity of the conclusions.

Acknowledgements This research was partially supported by Spanish Government grant [SEJ2006-27587-E/SOCI, DEP2015- 66069-P, and PSI2015-71947-REDT], as well as was partially supported by Generalitat de Catalunya [2014 SGR 971].

References

Anguera MT (2003) Observational methods (general). In: Fernández-Ballesteros R (ed) Encyclopedia of behavioral assessment, vol 2. Sage. London, UK, pp 632–637

Anguera MT, Camerino O, Castañer M, Sánchez-Algarra P, Onwuegbuzie AJ (2017) The specificity of observational studies in physical activity and sports sciences: moving forward in mixed methods research and proposals for achieving quantitative and qualitative symmetry. Front Psychol 8:2196. https://doi.org/10.3389/fpsyg.2017.02196

Bakker AB, Daniels K (eds) (2013) A day in the life of a happy worker. Psychology Press, Hove Sussex

Bell IH, Lim MH, Rossell SL, Thomas N (2017) Ecological momentary assessment and intervention in the treatment of psychotic disorders: A systematic review. Psychiatric services. https://doi.org/10.1176/appi.ps.201600523

Blair AJ, Leakey PN, Rust SR, Shaw S, Benison D, Sandler DA (1999) Locus of control and mood following myocardial infarction. Coron Health Care 3:140–144

Bolger N, De Longis A, Kessler RC, Schilling EA (1989) Effects of daily stress on negative mood. J Pers Soc Psychol 57(5):808–818

Bradburn NM (1969) The structure of psychological well-being. Aldine, Chicago

Bradley MM, Lang PJ (1994) Measuring emotion: The self-assessment manikin and the semantic differential. J Behav Ther Exp Psychiatry 25:49–59

Brunswik E (1944) Distal focusing of perception: Size constancy in a representative sample of situations. Psychol Monogr 56:1–49

Brunswik E (1956) Perception and the representative design of experiments, 2nd edn. University of California Press, Berkeley

Byrk AS, Raudenbush SW (2002) Hierarchical linear models: Applications and data analysis methods, 2nd edn. Sage Publications, Newbury Park

Clark LA, Watson D (1988) Mood and the mundane: Relations between daily life events and self-reported mood. Journal of Personality and Social Psychology 54(2):296–308

Clark LA, Watson D, Leeka J (1988) Diurnal variation in the positive affects. Motiv Emot 13(3):205–234

Cojuharenco I, Ryvkin D (2008) Peak-end rule versus average utility: How utility aggregation affects evaluation of experiences. J Math Psychol 52:326–335

Conner TS, Mehl MR (2015) Ambulatory assessment methods for studying everyday life. In: Scott R, Kosslyn S, Pinkerton N (eds) Emerging trends in the social and behavioral sciences, Wiley, Hoboken

Conner TS, Tennen H, Fleeson W, Barrett LF (2009) Experience sampling methods: A modern idiographic approach to personality research. Soc Pers Psychol Compass 3(3):292–313

Cordier R, Brown N, Chen YW, Wilkes-Gillan S, Falkmer T (2014) Piloting the use of experience sampling method to investigate the everyday social experiences of children with asperger syndrome/high functioning autism. Dev Neurorehabilitation 19(2):103–110

Creswell JW, Plano Clark VL (2011) Designing and conducting mixed methods research, 2nd edn. Sage, Thousand Oaks

Csikszentmihalyi M (1990) Flow: The psychology of optimal experience. HarperCollins, New York

Cunningham MR (1979) Weather, mood, and helping behavior: Quasi experiments with the sunshine samaratin. J Pers Soc Psychol 37(11):1947–1956

David JP, Green PJ, Martin R, Suls J (1997) Differential roles of neuroticism, extraversion, and event desirability for mood in daily life: An integrative model of top-down and bottom-up influences. J Pers Soc Psychol 73(1):149–159

Denissen JJA, Butalid L, Penke L, van Aken MAG (2008) The effects of weather on daily mood: A multilevel approach. Emotion 8(5):662–667

Diener E, Seligman MEP (2004) Beyond money: Towards an economy of well-being. Psychol Sci Public Interes 5(1):1–31

Diener E, Oishi S, Lucas RE (2003) Personality, culture, and subjective well-being: Emotional and cognitive evaluations of life. Ann Rev Psychol 54:403–425

Doherty ST, Lemieux CJ, Canally C (2014) Tracking human activity and well-being in natural environments using wearable sensors and experience sampling. Soc Sci Med 106:83–92

Dubad M, Winsper C, Meyer C, Livanou M, Marwaha S (2017) A systematic review of the psychometric properties, usability and clinical impacts of mobile mood-monitoring applications in young people. Psychol Med. https://doi.org/10.1017/S0033291717001659

Fahrenberg J, Myrtek M, Pawlik K, Perrez M (2007) Ambulatory assessment monitoring behavior in daily life settings: A behavioral-scientific challenge for psychology. Eur J Psychol Assess 23(4):206–213

Frederickson B, Kahneman D (1993) Duration neglect in retrospective evaluations of affective episodes. J Pers Soc Psychol 65:45–55

Frey BS, Stutzer A (2002) What can economists learn from happiness research? J Econ Lit 40:402–435

Goldstein H (1995) In: Arnold E (ed) Multilevel statistical models, vol 3, 2nd edn. Kendalls library of statistics. London

Goldstein KM (1972) Weather, mood, and internal-external control. Percept Mot Ski 35:786

Harmatz MG, Well AD, Overtree CE, Kawamura KY, Rosal M, Ockene IS (2000) Seasonal variation of depression and other moods: A longitudinal approach. J Biol Rhythm 15:344–350

Hektner JM, Schmidt JA, Csikszentmihalyi M (2007) Experience sampling method: Measuring the quality of everyday life. Sage Publications, Thousand Oaks

Hirshleifer D, Shumway T (2003) Good day sunshine: Stock returns and the weather. J Financ 58(3):1009–1032

Hogarth RM, Portell M, Cuxart A, Kolev GI (2011) Emotion and reason in everyday risk perception. J Behav Decis Mak 24:202–222. https://doi.org/10.1002/bdm.689

Hogarth RM (2005) The challenge of representative design in psychology and economics. J Econ Methodol 12:253–263

Hogarth RM, Portell M, Cuxart A (2007) What risks do people perceive in everyday life? A perspective gained from the experience sampling method (ESM). Risk Anal 27(6):1427–1439

Howarth E, Hoffman MS (1984) A multidimensional approach to the relationship between mood and weather. Br J Psychol 75:15–23

Kahneman D, Krueger AB, Schkade DA, Schwarz N, Stone AA (2004) A survey method of characterizing daily life experience: The day reconstruction method. Science 306(5702):1776–1780

Keller MC, Fredrickson BL, Ybarra O, Cté S, Johnson K, Mikels J, Conway A, Wager T (2005) A warm heart and a clear head: The contingent effects of weather in mood and cognition. Psychol Sci 16(9):724–731

Kemp S, Burt CDB, Furneaux L (2008) A test of the peak-end rule with extended autobiographical events. Mem Cogn 36(1):132–138

Klonowicz T (2001) Discontented people: Reactivity and locus of control as determinants of subjective well-being. Eur J Pers 15:29–47

Larsen RJ, Fredrickson BL (1999) Measurement issues in emotion research. In: Kahneman D, Diener E, Schwarz N (eds) Well-being: foundations of hedonic psychology. Russell Sage, New York, pp 40–60

Larsen RJ, Kasimatis M (1990) Individual differences in entrainment of mood to the weekly calendar. J Pers Soc Psychol 58(1):164–171

Liao Y, Skelton K, Dunton G, Bruening M (2016) A systematic review of methods and procedures used in ecological momentary assessments of diet and physical activity research in youth: An adapted strobe checklist for reporting ema studies (cremas). J Med Internet Res 18(6):151 https://doi.org/10.2196/jmir.4954

Longford NT (1993) Random coefficient models. Oxford University Press, Oxford

Magnusson M, Burgoon J, Casarrubea M (eds) (2016) Discovering hidden temporal patterns in behavior interaction. Springer, New York

Mehl MR, Conner TS (2012) Handbook of research methods for studying daily life. The Guilford Press, New York

Monk T, Fookson J, Moline M, Pollak C (1985) Diurnal variation in mood and performance in a time-isolated environment. Chronobiol Int 2(3):185–193

Moskowitz DS, Russell JJ, Sadikaj G, Sutton R (2009) Measuring people intensively. Can Psychol 50(3):131–140

Murray G, Allen NB, Trinder J (2002) Mood and the circadian system: Investigation of a circadian component in positive affect. Chronobiol Int 19(6):1151–1169

Murray G, Nicholas CL, Kleiman J, Dwyer R, Carrington MJ, Allen NB, Trinder J (2009) Nature's clocks and human mood: The circadian system modulates reward motivation. Emotion 9(5):705–716

Oishi S, Diener E, Choi DW, Kim-Prieto C, Choi I (2007) The dynamics of daily events and well-being across cultures: When less is more. J Pers Soc Psychol 93(4):685–698

Os J, Verhagen S, Marsman A, Peeters F, Bak M, Marcelis M, Glksz S (2017) The experience sampling method as an mhealth tool to support self-monitoring, self-insight, and personalized health care in clinical practice. Depress Anxiety 34(6):481–493

Parrott WG, Sabini J (1990) Mood and memory under natural conditions: Evidence for mood incongruent recall. J Pers Soc Psychol 59(2):321–336

Peeters F, Berkhof J, Delespaul P, Rottenberg J, Nicolson NA (2006) Diurnal mood variation in major depressive disorder. Emotion 6(3):383–391

Portell M, Anguera MT, Chacón-Moscoso S, Sanduvete-Chaves S (2015a) Guidelines for reporting evaluations based on observational methodology. Psicothema 27(3):283–289

Portell M, Anguera MT, Hernández-Mendo A, Jonsson GK (2015c) Quantifying biopsychosocial aspects in everyday contexts: an integrative methodological approach from the behavioral sciences. Psychol Res Behav Manag 8:153–160

Portell M, Anguera MT, Hernández-Mendo A, Jonsson GK (2015b) The legacy of brunswik's representative design in the 21st century: methodological innovations for studying everyday life. The Brunswik Society Newsletter. http://www.brunswik.org/newsletters/2015news.pdf. Accessed Dec 2015

Rind B (1966) Effects of beliefs about weather conditions on tipping. J Appl Soc Psychol 26(2):137–147

Rind B, Strohmetz D (2001) Effects of belief about future weather conditions on restaurant tipping. J Appl Soc Psychol 31(10):2160–2164

Rossi AS, Rossi PE (1977) Body time and social time: Mood patterns by menstrual cycle phase and day of the week. Soc Sci Res 6:273–308

Rotter JB (1966) Generalized expectancies for internal versus external control of reinforcement. Psychol Monogr 80(1, Whole No. 609)

Ruys KI, Stapel DA (2008) The secret life of emotions. Psychol Sci 19:385–391

Sanders JL, Brizzolara MS (1982) Relationships between weather and mood. J Gen Psychol 107:155–156

Sandstrom GM, Lathia N, Mascolo C, Rentfrow PJ (2017) Putting mood in context: Using smartphones to examine how people feel in different locations. J Res Pers 69:96–101

Santangelo P, Bohus M, Ebner-Priemer UW (2014) Ecological momentary assessment in borderline personality disorder: A review of recent findings and methodological challenges. J Pers Disord 28(4):555–576

Saunders EM (1993) Stock prices and wall street weather. Am Econ Rev 83:1337–1345

Schkade DA, Kahneman D (1998) Does living in california make people happy? A Focus Illusion JudgmS Life Satisf 9(5):340–346

Schwartz N, Clore GL (1983) Mood, misattribution, and judgments of well-being: Informative and directive functions of affective states. J Pers Soc Psychol 45(3):513–523

Shiffman S, Stone AA, Hufford MR (2008) Ecological momentary assessment. Ann Rev Clin Psychol 4:1–32

Silvia PJ, Kwapil TR, Walsh MA, Myin-Germeys I (2014) Planned missing data designs in experience sampling research: Monte carlo simulations of efficient designs for assessing within-person constructs. Behav Res Methods 46(1):41–54

Simonsohn U (2007) Clouds make nerds look good: Field evidence of the impact of incidental factors on decision making. J Behav Decis Mak 20(2):143–152

Smith TW (1979) Happiness: Time trends, seasonal variations, intersurvey differences, and other mysteries. Soc Psychol Q 42(1):18–30

Strack F, Argyle M, Schwartz N (eds) (1991) Subjective wellbeing: An interdisciplinary perspective. Pergamon Press, Oxford

Stone AA, Hedges SM, Neale JM, Satin MS (1985) Prospective and cross-sectional mood reports offer no evidence of a blue monday phenomenon. J Pers Soc Psychol 49(1):129–134

Stone AA, Smyth JM, Pickering T, Schwartz J (1996) Daily mood variability: Form of diurnal patterns and determinants of diurnal patterns. J Appl Soc Psychol 26(14):1286–1305

Stone AA, Schwartz JE, Schwartz N, Schkade D, Krueger A, Kahneman D (2006) A population approach to the study of emotion: Diurnal rhythms of a working day examined with the day reconstruction method. Emotion 6(1):139–149

Strack F, Martin LL, Schwartz N (1988) Priming and communication; social determinants of information use in judgments of life satisfaction. Eur J Soc Psychol 18(5):429–442

Thayer RE (1987) Problem perception, optimism, and related states as a function of the time of day (diurnal rhythm) and moderate exercise: Two arousal systems in interaction. Motiv Emot 11(1):19–36

Trombley MA (1997) Stock prices and wall street weather: Additional evidence. Q J Bus Econ 36:11–21

Trull TJ, Ebner-Priemer U (2014) The role of ambulatory assessment in psychological science. Curr Dir Psychol Sci 23(6):466–470

Watson D, Tellegen A (1985) Toward a consensual structure of mood. Psychol Bull 98:219–235

Watson D, Wiese D, Vaidya J, Tellegen A (1999) The two general activation systems of affect: Structural findings, evolutionary considerations, and psychobiological evidence. J Pers Soc Psychol 76(5):820–838

Wen CKF, Schneider S, Stone AA, Spruijt-Metz D (2017) Compliance with mobile ecological momentary assessment protocols in children and adolescents: A systematic review and meta-analysis. J Med Internet Res 19(4):132. https://doi.org/10.2196/jmir.6641

Wood C, Magnello M (1992) Diurnal changes in perceptions of energy and mood. J R Soc Med 85:191–194

Wood C, Magnello M, Sharpe MC (1992) Fluctuations in perceived energy and mood among patients with chronic fatigue syndrome. J R Soc Med 85:195–198

Volmer J, Fritsche A (2016) Daily negative work events and employees' physiological and psychological reactions. Front Psychol 7:1711. https://doi.org/10.3389/fpsyg.2016.01711

Zembylas M, Schutz PA (eds) (2016) Methodological advances in research on emotion and education. Springer, New York

Chapter 5
The Observer Experiment: A View of the Dynamics of Multimodal Interaction

Ghazaleh Esfandiari-Baiat, Laszlo Hunyadi and Anna Esposito

Abstract Here we present and also apply the "observer gaze" methodology in order to study the structural organization of multimodal, face-to-face conversations. Tracking the gaze of a third party observer can provide an online and implicit method for investigating the dynamics of interactions and the relevant features that capture the observer's attention according to the amount of information the observer has available. The way an outsider visually perceives the interactional exchanges may also suggest a better modeling of human-human communication for human-machine communication system. With these goals, ten observers participated in the present study, all being native speakers of Persian. The observers watched a short clip of a dyadic, spontaneous conversation in Persian. Using an eye tracker, the observers' gaze was recorded in two different watching situations, namely: audio-visual, and visual-only. Afterwards, using ELAN, the observers' gaze data were temporally aligned with verbal and nonverbal behavior of the interactants (such as head movements, hand gestures, body shift, facial expressions and turn-taking). First, the observers' visual attention, measured as the amount of fixation time on the speaker and listener, was described across conditions. Second, the observers' gaze was analyzed in presence of turn-transitions and we showed that they were able to a large extent to anticipate turns and also follow them depending on the conditions. In summary, the observers' gazing behavior was affected by the conditions.

G. Esfandiari-Baiat (✉) · L. Hunyadi
Department of General and Applied Linguistics, University of Debrecen,
Debrecen, Hungary
e-mail: esfandiari.gh@gmail.com

L. Hunyadi
e-mail: hunyadi@ling.arts.unideb.hu

A. Esposito
Department of Psychology, Seconda Universitá di Napoli, 34 Viale Ellittico,
81100 Caserta, Italy
e-mail: iiass.annaesp@tin.it

5.1 Introduction

Humans, considered as social beings, are involved in all kinds of interactions on a daily basis and for several individual and social purposes. No matter if it is a simple greeting with a friend, or a business discussion at a meeting, a visit to the doctor, or some kind of shopping event, all of these are accomplished through interactional exchanges. It is only through such interactions that humans cooperate and move on in their daily life. Understanding human behaviors is not possible without studying the dynamics and structures of such inter-personal exchanges. Regardless of how different the purposes of these interactions are, they are dynamically implemented through a coordinated turn-taking system that is a universal communicative feature.

In effect, communication is carried out through speakers' nearly continuous turns, with minimal gaps, averaging 200 ms or less in face-to-face conversation (Brady 1968), 700 ms over the phone (Jaffe and Feldstein 1970) and less than 5% of speech overlaps (Levinson 1983). This general pattern of minimal gaps-minimal overlaps in turns' shifts is considered universal and was observed across many cultures (Stivers et al. 2009; Tice and Henetz 2011). Such efficiency raises the question of how actors involved in a conversation can manage this precise timing. A study by Sacks et al. (1974) shows that the language production system is quite slow. Therefore, inter-speaker gaps (silences) are normally too brief for listeners to rely on as a cue to start their response. A single word requires 600 ms from conception to articulatory output (Indefrey and Levelt 2004; Levelt 1989) and multiword utterances take considerably longer (Schnurr et al. 2006; Jescheniak et al. 2003; Magyari et al. 2014). Sacks et al. suggested that while comprehending the current utterance listeners are simultaneously able to track indications for predicting their ongoing turn using information about the syntactic, propositional, and intonational structure of utterances. Exploiting these cues, listeners are able to predict their turns and plan their responses to a high accuracy (Casillas and Frank 2012).

Since the publication of experimental results by Sacks et al., researchers have investigated turn-taking mechanisms and attempted to identify the verbal and non-verbal cues that make possible this accurate coordination. These studies exploited only audio or audio and video corpora of dyadic conversations to identify the most common informative verbal and nonverbal cues for turn-completion. They found that syntactic cues as well as intonational features and some notion of pragmatic or action completion define transitions relevance places in conversations (Ford and Thompson 1996, p. 171). Caspers (2003) found that pitch accents and boundary tones also contribute to inform the actors of turn completions. Abuczki (2012) reported that the most relevant cue for turn taking is the gaze behavior of the actors. However, other authors pointed out that no single cue in isolation can act as the most relevant in detecting turn taking given the multimodal nature of communication (Norris 2004; Stivers and Sidnell 2005; Hunyadi 2011) and some have debated their effectiveness because these cues might come too late to play any decisive role for listeners to make use of them (Magyari et al. 2014). All of these studies are concentrated on the actors and on the cues they exploit to predict the upcoming turn. To our knowledge there are only two studies that investigate turn completion cues from the point of view of the observer. The first study was made by De Ruiter et al. (2006) in order to find cues that were more informative in the process of turn-boundary anticipation. They created an on-line experiment, manipulating *"the presence of symbolic (lexico-syntactic) content and intonational contour of utterances recorded in natural conversations"* (De Ruiter et al. 2006, p. 515). They asked Dutch speakers to listen to spontaneous speech fragments of conversations which were phonetically manipulated (lexico-syntactic content and intonational contour were either inserted in or removed from the utterance) and press a button at the moment they anticipated the speaker would finish her utterance. Their results indicated that the lexico-syntactic content was necessary for anticipating the utterance completion, and hence regulating conversational turn-taking. This was not the case for intonational contour which was found insufficient for end-of-turn projections.

The second study, from which the present paper takes inspiration, was carried out by Tice and Henetz (2011). They explored eye tracking as a possible alternative implicit method for measuring online turn processing compared to the explicit experimental method used by Due Ruiter and colleagues. In their study, Tice and Henetz (2011) recorded the eye movements of subjects while they were watching two short "split-screen" dialogues from a film Mean Girls (Paramount Pictures 2004), where two girls were speaking over the phone (the observers were able to see their faces). They found that observer gaze behaviors reliably anticipate next speaker turns. However, these results may have been affected by a methodological issue related to the experimental setting. The observers did not have any other choice than to look at the faces of the speakers, as the body and the body gestures were not present. In addition, since the conversation was taken from a well-known film, the observers might have already seen and heard the conversation and this could influence their turn taking predictions.

The present study attempts to overcome the above-mentioned limitations. Firstly, as can be seen in Fig. 5.1, the complete scene of the conversation (the full body of the actors and also the physical context of the conversation) was included in the video-clip that the observers watched. This allowed us to take into account all the observers' gaze behaviors. Secondly, the actors involved in the conversation were completely and equally unfamiliar to the observers, making their predictions more realistic.

The aim of the current study is to present an evaluation of the interaction flow dynamics through the eyes of observers viewing a conversation from the outside in two different conditions, namely audio-visual and visual-only. We wanted to analyze and compare their gazing behavior in order to identify valuable cues about how they process conversations. The research questions to be answered in the study are the following:

1. Where do the observers look? Do they mainly focus on the actors?
2. Do they pay equal attention to the actors or is one more interesting to them?
3. Do they also pay any visual attention to the listener (the non-speaking actor)?
4. Does the observers' gazing behavior differ as the result of changing conditions?

5. Do they pay any attention to the gestures produced by the actors?
6. Do the observers visually anticipate or follow the speakers' turns in the conversation?

5.2   Materials and Methods

5.2.1   Participants

Ten subjects (five females and five males) participated in this case study. They were all aged 20–24, studying at Debrecen University (HU) as a foreign student. Their mother tongue was Persian. They all participated in the audio-visual and visual-only setup and were unfamiliar with the actors involved in the watched conversation. Their participation in the experiment was voluntarily.

5.2.2   Materials

The observed conversation was recorded in the HuComTech (Hunyadi et al. 2012) sound-proof studio of the University of Debrecen. It was a spontaneous, face-to-face dialogue between two native Persian speaking students from Iran, studying as foreign students at the University of Debrecen. They sat opposite to each other during the conversation and the only instruction given to them was to talk about their experiences as a foreign student in Hungary. They were good friends and the dialogue was informal and lasted about 20 min. The conversation was recorded with two kardioid microphones (AT2035) and an external sound card to separate the actors' channels. Three cameras were used, the first camera capturing actor A only (the boy), the second capturing actor B only (the girl), and the third capturing both of them with a view of the whole conversation. Figure 5.1 shows a screenshot of the video clip which was played for the observers during the experiment.

The conversation was subsequently transcribed and annotated on various tiers using the ELAN (Brugman and Russel 2004) annotation tool. Head movements, hand gestures, facial expressions and turn boundaries were marked for both actors. Later, the observers' gaze data were imported into ELAN and synchronized with the rest of the annotations.

5.2.3   Procedure

The observers' gazes were first calibrated using the Viewpoint remote eye-tracker auto-calibration system (a 16-point calibration was used). The observers were seated in front of a monitor during the experiment while the first two minutes of the

Fig. 5.1 The conversational scene

conversation was played on the monitor and their gaze movements were simulta-neously recorded. In order to find out where the observers were looking, four regions of interest (henceforth: ROIs) were defined in order to facilitate data analysis, each representing a specific physical region of the conversational venue. Regions one (R1) and three (R3) are associated with the face of the actors A and B, respectively. Re-gions two (R2) and four (R4) present the upper body part (including the estimated space for the hand movements) of actors A and B, respectively.

The observers were not informed about the goal of the experiment. In order to make their observations as natural as possible, no specific instructions were given to them. They were only told that they would watch short clips of dyadic conversations and would be asked about the content of the conversation and the personalities of the interactants. The experiment was carried out in two different conditions that provided the opportunity to compare the gazing behavior of individuals in the presence or absence of different modalities. In the first condition (visual-only), the observers could only see but not hear the conversation. In the second condition (audio-visual setup), the same observer could both see and also hear the conversation. In order to avoid learning effects, after seeing the visual-only experimental video, the observers watched two similar distracting videos, the first mute and the second complete Persian dyadic conversation, and then they watched the complete video of the experimental conversation.

The reason for having different conditions was to control the experiment for factors that might influence how the observers process turn-taking. In the first condition the auditory modality was fully blocked. In this condition the observers did not have access to lexico-syntactic and prosodic information. They only had access to visual features. However, in the second condition the observers had full access to visual, lexico-syntactic, and prosodic information. A comparison of the gazing behavior of the observers in these two conditions can underline the importance of modalities and visual, lexico-syntactic or prosodic cues mostly exploited for predicting turn takings.

5.3   Results and Discussion

5.3.1   General Overview of Gaze Fixation Time

Now we will provide a general overview of the data. The design of the experiment was aimed at accurately following the visual attention of the observer in both conditions so that based on the data received we could compare the gazing behavior in the audio-visual and video-only setups. We wanted to know whether the change of setting could affect gaze fixation time in the ROIs. Table 5.1 shows the average gaze fixation (duration) in each ROI and also the amount of time which was spent outside the regions in both conditions. These data were calculated independently of what was happening in the conversation.

Table 5.1 shows the average time (in ms) spent for fixation in each region of interest and outside them by the observers. It also shows this amount of time in percentage terms, computed as the ratio of the time the observer spent in fixating at a given region over the total time of the conversation. The results tell us that, comparing the two conditions, the gaze stayed within the ROIs for 81.3% of the time in audio-visual condition and for 78.25% of the time in the visual-only condition, for 79.8% in average for the two conditions together. This suggests the most relevant features capturing the visual attention of the observers, independently from the conditions, were the faces (facial expressions and head movements) and the body movements of the actors. Although there were several objects in the scene to look at, they were mostly focused on the actors. As can be deduced from the data in Table 5.1, the gazing behavior of the observers did not differ dramatically in the two conditions, meaning that their visual attention was not affected by the difference in the experimental setups.

In both conditions, the observers mostly focused on the faces (R1 and R3) of the actors. They spent much less time in regions R2 and R4 which were associated with the actors' hand gestures and body movements. However, more time was spent on these regions in the sound-off condition, indicating that in the absence of one modality (auditory channel), observers were relied more on other modalities (visual channel) in order to better understand the conversation. This is illustrated in Fig. 5.2, which shows the gaze fixation time (as a percentage) in each condition, each region of interest, and each actor.

Table 5.1 Average gaze fixation time in and outside the ROIs across conditions

Conditions	R1	%	R2	%	R3	%	R4	%	Outside ROI	%
Audio-visual	51728.8	44.2	3268.8	2.8	38230.4	32.7	1884.4	1.6	21887.5	18.7
Visual-only	39975	34.2	7053	6	40863.7	34.9	3685.8	3.15	25422.5	21.7

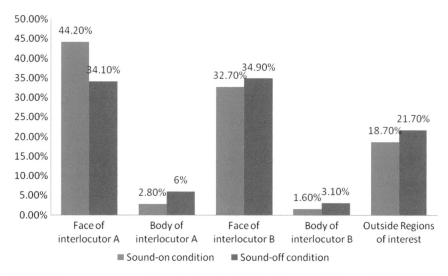

Fig. 5.2 Average percentage of gaze fixation time distributed over conditions and ROIs

5.3.2   *Proportion of Gaze on Speaker and Listener*

While the previous section provided a general overview of the average gaze fixation time in all ROIs regardless of who was speaking, here we analyze the proportion of gaze fixation on the current speaker and listener (non-speaking actor). We wanted to know where the observers look when actor *A* or *B* was speaking to see whether the observers follow the speaker all the time (100%) or the listener is also of interest to them. Tables 5.2 and 5.3 list the related gaze data. Table 5.2 gives the observers' gaze fixation time on the actors in both conditions when actor *A* is currently speaking and actor *B* is the listener. Table 5.3 lists the same data for the situation where actor *B* is speaking and actor *A* is listening. The time is given in terms of ms and percentages. In Table 5.2 the percentages for the column *SA* are computed as the ratio of the total observer gaze fixation time on actor A over the time actor *A* is speaking. In the

Table 5.2 Average gaze fixation time in the ROIs when interlocutor A is talking

Conditions	R1	%	R2	%	SA	R3	%	R4	%	SB
Audio-visual	3880	54.7	2158	3	57.7	14436	20.3	2039	2	22.3
Visual-only	31232	44	5743	8.1	51.1	15511	21.8	2174	3	24.8

Table 5.3 Average gaze fixation time in the ROIs when interlocutor B is talking

Conditions	R1	%	R2	%	SA	R3	%	R4	%	SB
Audio-visual	9421	22.5	586	1.2	23.7	22055.2	52.8	1635.4	3.9	56.7
Visual-only	10638.9	25.5	934.7	2.1	27.6	22175.3	53.1	1487.7	2.8	55.9

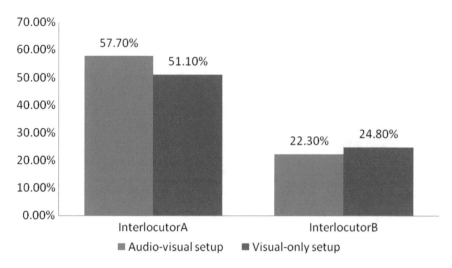

Fig. 5.3 Average gaze fixation time when interlocutor A is speaking and interlocutor B is listening in both conditions

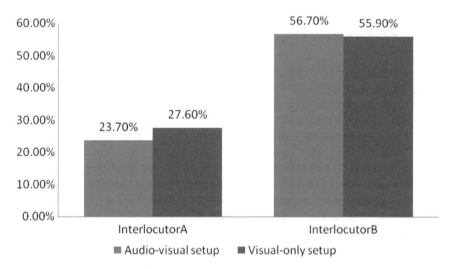

Fig. 5.4 Average gaze fixation time when interlocutor B is speaking and interlocutor A is listening in both conditions

column SB the percentages are computed as the ratio of the observer gaze fixation time on actor B over the time actor A is speaking. In Table 5.3 the situation is reversed. Figures 5.3 and 5.4 present the same data as histograms.

Based on the data reported in Tables 5.2 and 5.3, independent of the condition, the observers' gaze was mostly on the current speaker. Similar results were also reported in a previous study by Casillas and Frank (2012). They claimed that the observers glanced at the listener around 20% of the time. Our study shows that gaze fixation time on the speaker and listener is affected by the conditions. These results are in accord with those of Casillas and Frank (2012) for the audio-visual and the visual-only conditions. In general, although the visual attention of the observers were mostly on the current speaker, the time they spent watching the listener was also consistent, suggesting that the listener plays an important role as an active element in constructing and coordinating the conversation. The verbal (i.e. short vocal feedbacks) and nonverbal (i.e. head nods and smiles) behaviors performed by the listener may be the reason of attracting the visual attention of the observers. The highest percentage of gaze on the speaker was as expected, in the audio-visual setup, because in this condition the observer could hear and understand the conversation while the lowest percentage was in the visual-only setup.

5.3.3 Trajectory of Gaze

The two preceding sections described evidence related to the observers' visual attention and their gazing behavior towards the speaker and listener. Now we will present the trajectory of the observers' gaze. We wanted to learn what the most frequent gaze's trajectories were from one ROI to another (Fig. 5.5).

As shown in Table 5.4, the most frequently observed sequential pattern of gaze trajectory is R1–R2 in both the audio-visual and the visual-only conditions meaning that a gaze hit in region one was followed by a gaze hit in region two. Such pattern

Fig. 5.5 The variable match search in ELAN

Table 5.4 Sequential gaze patterns across conditions

Conditions	Number of variables	Count	Percentage	Sequential gaze pattern
Audio-visual setup	(x, y)	35	32.4%	R1–R2
	(x, y, z)	20 18.7%	R1–R2–R1	
Visual-only setup	(x, y)	35	24%	R1–R2
	(x, y, z)	24	16.7%	R2–R1–R2

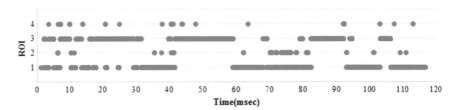

Fig. 5.6 A sample of a gaze trajectory in the audio-visual setup

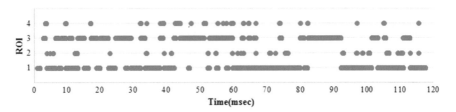

Fig. 5.7 A sample of a gaze trajectory in the visual-only setup

indicates that the observers were mostly screening actor A by consecutively looking at his face followed by his body. This might be explained by the fact that actor A was the more dynamic participant, both verbally and non-verbally, in the conversation. Figure 5.6 visualizes a sample of the gaze trajectory in the audio-visual setup and Fig. 5.7 shows the same data for visual-only condition. Each dot represents a single gaze hit in different regions of interest.

Comparing the two conditions, the observers perform more gaze hits in sound-off condition, especially in the regions which are related to the body parts of the actors. This might be because the observers do not have access to the verbal content of the conversation, so they rely more on the gestures as a source of information to find out what is going on. In addition, in this condition the gaze points are more scattered over the ROI compared to the sound-on one. Also, the gaze fixations seem shorter in the sound-off compared to the sound-on condition. Figures 5.6 and 5.7 illustrate the gaze trajectory for one observer in two different conditions.

5.3.4 Gazing Behavior at Turn-Transition Points

After describing the most frequently observed sequential patterns of gaze, this section presents the data related to the gazing behavior of the observers at turn-transition points. We wanted to know where the observers look in the surrounding environment of turn-transitions. Do they visually anticipate the turn-transition points or do they follow the transitions with their gaze? When an ongoing conversation nears a point of turn transition, observers may look to the next speaker in advance, that is, before the current speaker has stopped speaking. Such early gaze movements towards the upcoming speaker might be indicative of turn-end anticipation. Tracking the gaze in both conditions, we wanted to know if the observers had such an anticipatory gaze.

There were 19 conversational turns annotated altogether, from which 10 turns were associated with actor A and 9 turns with actor B. The conversational turns were annotated as periods in time during which only one speaker was talking (disregarding back-channels produced by the listener). Overlapping speech was also marked separately. Using the multiple layer search in ELAN, the gazing behavior of the observers were analyzed during turn transitions in different conditions. The search results were also checked manually in the annotation file. In order to investigate the gazing behavior of the observers during a turn transition, 300 ms before and after the transition was analyzed (referred to as the gaze window). Comparing the gaze window before and after the transition, we found two different types of gazing behavior for turn transitions. In the first type, the gaze was in a different ROI before and after the transition hence, the gaze shifted from one region to another as the turn was transferred. In the second type, the gaze stayed in the same ROI before and after the transition and meant there were no gaze shifts in the immediate part of the turn transition. In the next section, each condition will be described separately.

As it is seen in Fig. 5.8, in both conditions, only around 12% of the turn transitions on average were not visually noticed by the observers. These conversational turns were short and located at the very beginning of the conversation. In both conditions there were more instances where the observer did not perform a gaze shift at turn-transition points, meaning that the turn transition occurred within one gazing block (the gaze staying in the same ROI before and after the transition). However, there

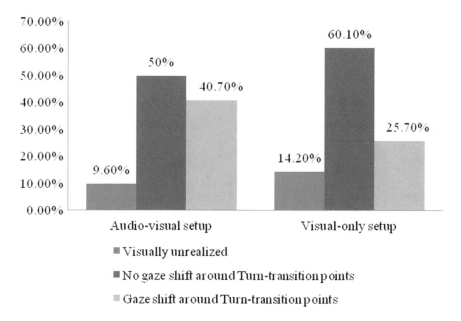

Fig. 5.8 Gazing behavior of observers around turn transition points across conditions

were also instances where a gaze shift was present at the transition points. The highest percentage (40.7%) where a gaze shift occurred was in the audio-visual condition where the observers had full access to the content of the conversation and the lowest percentage (25.7%) is observed in the visual-only condition where the least source of information was present for the observer.

Looking at the transition points in the audio-visual setup, there were 10 instances (52.6%) where the observer moved her gaze towards the upcoming speaker in advance, before s/he had started speaking (on average 400 ms before the transition). Such gaze shifts might be associated with the observer anticipating the point of turn transition. In addition, there were 4 instances (21%) where the observer's gaze moved towards the actor only after s/he had already started speaking. In these cases the gaze stays on the previous speaker at the beginning of the turn and only after a while (on average 1585 ms) moves to the current speaker. There were 5 turns (26.3%) which were not visually followed by observers, showing no gazing activity around these turns.

In the visual-only setup, there were 7 cases (36.8%) of gaze anticipating turn transition (on average 610 ms before the transition) and there were 5 cases (26.3%) where the observer followed the speaker after s/he had started speaking (on average 1314 ms after the transition). For a summary see Fig. 5.9.

Regardless of the conditions, the highest percentage is related to the observer gaze anticipating the turn transition not following it. The gaze shifts from the speaking actor occur sometime before the turn transitions. In most cases, the observer shifts her gaze away from the current speaker to the next speaker as the turn gets close to the end and remains there (at the same ROI) as the turn-transition occurs. Based on these results, we can say that the observers were more inclined to anticipate turn transition rather than following it. However, the rate of turn-transition anticipation

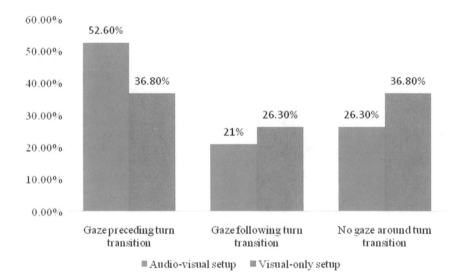

Fig. 5.9 Gaze anticipating or following turn-transitions

Table 5.5 Alignment of the observer's gaze with the non-verbal behaviors in different conditions

Conditions	Alignment of gaze with head movements (%)	Alignment of gaze with hand gestures (%)	Alignment of gaze with facial expressions (%)
Audio-visual setup	21.7	5.1	4.2
Visual-only setup	15.9	17.9	16.7

is higher in the sound-on condition where the observer can hear and also understand the language. The percentage of anticipation drops in the sound-off condition. These results highlight the importance of the lexico-syntactic content of a conversational turn for helping the observers to project the moment of its completion, and hence regulating conversational turn-taking. These results are in accord with a previous study carried out by De Ruiter and his colleagues on the end-of-turn projection. However, the anticipation rate observed in this study is lower than what was reported by Tice and Henetz in their study on turn-boundary anticipation. They reported that in most of the turn transitions the observers, both adults and children, performed anticipatory gazing behavior.

The conversational turns that were anticipated by the observers were mainly questions and answers (adjacency pairs) belonging to the start of the conversation. Looking at the video, these turns were also accompanied by strong non-verbal behaviors such as smiling, laughing and head movements of the actors. One or a combination of these cues might have also given the observer the cue that the current turn would finish soon. Table 5.5 shows an example of the alignment of an observer's gaze with the non-verbal behaviors of the actors. It shows the number of instances where the observer paid visual attention to a type of non-verbal behavior of the actoractors.

In the sound-on condition, the observer paid more attention to the head movements (nods, tilts, jerk, waggle and shake) compared to the other non-verbal behaviors such as hand movements and facial expressions. In this condition, the observer pays less visual attention to the non-verbal behaviors. Such gazing behavior might be due to the fact that she has access to the semantic content of the conversation. Since the observer hears and understands the conversation through the auditory modality she finds it unnecessary to focus on gestures. In the sound-off condition, there is more visual attention on the non-verbal behaviors especially on the hand movements. We assume that since the observer lacks one modality to rely on in this condition, tries to rely more on the still available other modality. By visually focusing more on the non-verbal behaviors of the actors, the observer is trying to understand what is going on in the conversation.

5.4   Conclusion

In this study we used gaze tracking as a means to study the structural organization of face-to-face interaction. We wanted to know how an observer who is not an active part of an ongoing conversation perceives its organizational structure. For this reason, observers were asked to watch a short clip from a conversation while their gaze was recorded. In order to control the effecting factors, the experiment was designed using two conditions: audio-visual and visual-only. We analyzed the observer's gaze in both conditions. Regardless of the conditions, the observers were mainly focused on the actors involved in the conversation. They spent most of their time (more than 85%) inside the ROI which were associated with the actors. More time was spent on the faces of the actors than on the other parts of their body capturing body movements and hand gestures. However, comparing the conditions, there was more visual attention recorded on the gestures in visual-only setup (more fixation on hand gestures).

The observers spent most of their time (more than 50%) on the actual speaker in both condition. However, they also glanced back at the listener (the non-speaking actor) quite often, especially during longer conversational turns. The fixation on the speaker was more constant in the sound-on condition compared to the other condition. In the sound-off condition the gaze was more scattered and the fixations were much shorter. Looking at the trajectory of gaze, the most frequent gaze sequence observed was R1 followed by R2 in both conditions (sound-on and sound-off) showing that the observers were more visually integrated with actor A who was more active both verbally and non-verbally. The observers were more visually integrated with turn transition in the audio-visual setup; because they had the highest number of gaze shifts during the transitions. However, the results indicates that regardless of the conditions, the observers more likely anticipated the turn transitions rather than following them. The highest rate of anticipation was in the audio-visual setup, highlighting the role of lexico-syntactic information for projecting turn transition points.

In the future we plan to continue this work by analyzing data from more Persian speaking observers. We would like to see whether our findings from this study are supported when analyzing data from further observers. We also plan to add a third condition (visual-prosody) to the study using Hungarian observers who are completely unfamiliar with Persian. The aim is to study the gazing behavior when both visual data and prosody are available and to compare with our results. It would also be interesting to ascertain both the effect of gender and cultural differences on the observers gazing behavior. Such studies could provide more insights into the cognitive processes underlying our ability to manage turn-taking in a dialogue. Looking at how and to what extent an outside observer is involved visually in a conversation should help us to gain a better understanding of conversational structure, and also get a better modeling of this unique communication system.

References

Abuczki Á (2012) A conversation analytical study on multimodal turn-giving cues. In: Esposito A, Esposito AM, Vinciarelli A, Hoffmann R, Müller VC (eds) Cognitive behavioural systems. Springer, Berlin, pp 335–342

Brady P (1968) A statistical analysis of on-off patterns in 16 conversations. Bell Syst Tech J 47:73–91

Brugman H, Russel A (2004) Annotating multimodal/multimedia resources with ELAN. In: Proceedings of LREC 2004, fourth international conference on language resources and evaluation

Casillas M, Frank M (2012) Cues to turn boundary prediction in adults and preschoolers. In: Brown-Schmidt S, Ginzburg J, Larsson S (eds) Proceedings of SemDial (SeineDial): the 16th workshop on the semantics and pragmatics of dialogue. Universit Paris-Diderot, Paris, pp 61–69

Caspers J (2003) Local speech melody as a limiting factor in the turn-taking system in Dutch. J Phon 31(2):251–276

De Ruiter JP, Mitterer H, Enfield N (2006) Projecting the end of a speaker's turn: a cognitive cornerstone of conversation. Language 82(3):515

Ford CE, Thompson SA (1996) Interactional units in conversation: syntactic, intonational, and pragmatic resources for the management of turns. Stud Interact Socioling 13:134–184

Hunyadi L (2011) Multimodal human-computer interaction technologies. Argumentum 7

Hunyadi L, Földesi A, Szekrényes I, Staudt A, Kiss H, Abuczki A, Bódog A (2012) Az ember-gép kommunikáció elméleti-technológiai modellje és nyelvtechnológiai vonatkozásai. [A theoretical-technological model of human-machine communication and its reference to language technology]. In: Kenesei I, Prószéky G, Váradi T (eds) Általános Nyelvészeti Tanulmányok XXIV. Akadémiai, Budapest

Indefrey P, Levelt WJM (2004) The spatial and temporal signatures of word production components. Cognition 92:101

Jaffe S, Feldstein S (1970) Rhythms of dialogue. Academic, New York

Jescheniak JD, Schriefers H, Hantsch A (2003) Utterance format affects phonological priming in the picture-word task: implications for models of phonological encoding in speech production. J Exp Psychol: Hum Percept Perform 29:441–454

Levelt WJM (1989) Speaking: from intention to articulation. MIT Press, Cambridge

Levinson S (1983) Pragmatics. Cambridge University Press, Cambridge

Magyari L, Bastiaansen MCM, De Ruiter JP, Levinson CS (2014) Early anticipation lies behind the speed of response in conversation. J Cogn Neurosci 26(11):2530–2539

Norris S (2004) Analyzing multimodal interaction: a methodological framework. Routledge, London

Sacks H, Schegloff EA, Jefferson G (1974) A simplest systematics for the organization of turn-taking for conversation. Language 50:696–735

Schnurr TT, Costa A, Caramazza A (2006) Planning at the phonological level during sentence production. J Psycholinguist Res 35:189–213

Stivers T, Sidnell J (2005) Introduction: multimodal interaction. Semiotica 156:1–20

Stivers T, Enfield NJ, Brown P, Englert C, Hayashi M, Heinemann T, Hoymann G, Rossano F, de Ruiter JP, Yoon KE, Levinson SC (2009) Universals and culturalvariation in turn-taking in conversation. Proc Natl Acad Sci 106(26):10,587–10,592

Tice M, Henetz T (2011) The eye gaze of 3rd party observers reflects turn-end boundary projection. In: Proceedings of the 15th workshop on the semantics and pragmatics of dialogue (SemDial 2011), Los Angeles, California, pp 204–205

Part III
Towards Application: Approches to the
Study of a Large Multimodal Corpus

Chapter 6
Uncertainty in Conversation: Its Formal Cues Across Modalities and Time

Laszlo Hunyadi

Abstract The topic of this chapter is a particular state of mind: uncertainty. One can experience uncertainty in many contexts and situations: in relation to events past, present and future, persons close and far, tasks, choices, measures, decisions desired or to be avoided, and much more. It is very personal, and therefore it has hardly innumerable forms of expression. Due to its personal nature, one may find it difficult to observe it by trying to associate it with just one or another single behavior, especially that it most often develops as a sequence of behavioral events over time. Based on the HuComTech multimodal corpus and using the computing environment Theme we show a complex of those individual events that, together, contribute to the perception of uncertainty. This complex is essentially temporal, in which patterns of behavior are formed from temporally adjacent or non-adjacent events including physically observable ones as well as those which are only interpretations of certain observations. The data are based on dialogues in Hungarian, but their multimodal analysis can be applied to other language environments as well.

6.1 Introduction

Uncertainty is one of those states of mind that may often be the most challenging to either perceive or interpret. Even though we would like to feel certain about what we experience or have knowledge of, this certainty is far from being fully predictable. Our everyday reasoning does not necessarily follow the deductive reasoning of categorical logic: our decisions or generalizations can be strongly affected by our beliefs or stereotypes, thus overriding the formalism of context-free syllogisms (Tversky and Kahneman 1973). Since contexts associated with an utterance can vary, our decisions

I dedicate this work to Aron, who will always be 27.
This chapter was supported by Grant NKFIH 116402

L. Hunyadi (✉)
Department of General and Applied Linguistics, University of Debrecen,
Debrecen, Hungary
e-mail: hunyadi@ling.arts.unideb.hu

© Springer Nature Switzerland AG 2020
L. Hunyadi and I. Szekrényes (eds.), *The Temporal Structure of Multimodal
Communication*, Intelligent Systems Reference Library 164,
https://doi.org/10.1007/978-3-030-22895-8_6

or generalizations based on it can also vary accordingly. It is especially noticeable in events of dynamically progressing social interaction: If the availability of more than one context is possible at a given time, the participants have to make a choice which context to select for a given act of interaction, be it a statement, question or answer. Obviously, the simplest kind of context is the one obtained from the linguistic form of the given interaction. A verbal utterance consists of words, their combination into phrases, sentences, and their formal, syntactic analysis mapped onto the corresponding semantic values of the lexical constituents yields the given basic context of the utterance. However, there can be a number of nonverbal parameters of an interaction that can override the above formally generated basic context. They are those that actualize the utterance for use in a given pragmatic situation, thus creating the pragmatic context necessary for the success of any social interaction (Bach and Harnish 1979; Sperber and Wilson 1986; Wilson and Sperber 2002, 2012). These parameters include the time and location of the given interaction, social norms, scenarios, beliefs and stereotypes, the social status and the psychological state of the actors, and many more (Hunyadi et al. 2010). In all, when making a decision in the conversation (what to ask, how to ask, what to answer, how to answer etc.), the actors follow a set of pragmatic requirements that effectively constrain them in making the appropriate choice. When doing so, however, they are facing a highly multivariate formula which allows for a number of parallel contexts and parallel choices to select from. This is where the perception of uncertainty may emerge: uncertainty can then be described as a process rather than some discrete state, a process of scanning through available choices. This process has two dimensions: a quantitative one—measured in time, from the start of scan to arriving at a decision (if any), and a qualitative one—constituting the evaluation of the possible choices. Whereas the temporal dimension is overtly observable, the content of the process of evaluation is not necessarily so. Even if the actor uses verbal expressions at the time of uncertainty, these expressions may or may not offer a reliable clue to the process of the evaluation. Their nonverbal expressions, on the other hand, are usually not specific enough to associate them with elements of this process. All that remains for the observer then is the perception of a state of uncertainty in general, unspecified in content but specified in time. Since content and time are considered independent parameters, in order to assign an interpretation to the perceived uncertainty, its duration alone will not be specific to the content. Since the emergence of uncertainty (as that of any other state of mind) is (at least partially) subject to the wider context of the given interaction, its interpretation requires the consideration of this larger context available for the observer, be it verbal or nonverbal; and within this context the duration of uncertainty, as the duration of any other constituent of the interaction may also have its place to consider.

The focus of this chapter will be the state of mind of uncertainty in social inter-actions, showing in what contexts it emerges. Since we understand uncertainty as a process, we will also search for its context as a process that gives rise to its emergence. There are all kinds of events taking place in an interaction, both verbal and nonver-bal. Just describing what events simultaneously co-occur with uncertainty will only show its wider environment in which it occurs, but not the cause of its emergence. Instead, we are going to search for temporally variable patterns in which uncertainty can be the result of another, preceding event or of a set of further preceding events, or where uncertainty results in the emergence of another, subsequent event or further subsequent events. Our work is based on data from the HuComTech multimodal corpus (Hunyadi 2011; Pápay et al. 2011), and the temporal patterns associated with uncertainty are identified using the Theme software environment (Magnusson 2000; Magnusson et al. 2016).

6.2 On the HuComTech Corpus: Purpose, Annotation and Data

The original purpose of building the corpus was to offer new insights into the multimodal complexities of human-human communication, such that could be further implemented in enhanced human-human interaction systems. The theory and practice of these systems are evolving rapidly (for an overview cf. Németh 2011; Hunyadi et al. 2012), demonstrated in more and more sophisticated applications, assisting the human user in a variety of fields and needs. Beyond their technical/technological capabilities, their actual effectiveness, however, relies greatly on how smooth an interaction with them is performed, i.e., how straightforward and effortless the user feels interacting with them. Ideally, the software interface behind the machine or application should notice, understand and properly respond to specific actions by the human and so potentially become an equal part of an event of human-like communication. With this aim in mind, our purpose was to design a set of conversations between humans that sufficiently model events representing the main ingredients of a conversation, such as turn management, questions, answers, agreement, attention, as well as elicitation of emotions. We made video recordings of two types of dialogues: formal (job interviews) and informal (free conversation, following a general scheme). Three cameras were used, two facing the speaker, one the agent. Both the speaker and the agent were sitting. There were 111 subjects (speakers, 60 male and 51 female) and 2 interviewers (agents, one male and one female), the same persons for both types of conversations with each subject. The cumulative recorded time was about 50 hours.

The annotation of data included both video and audio. Certain parameters were annotated multimodally (the annotator followed both the video and the audio), whereas others unimodally (only video or audio). The annotation of perceived emotions was done both multimodally and unimodally (audio only). Both physically observable values (gaze, movement of body parts) and interpretive values (emotions, pragmatic values) were annotated. Each category of behavior was annotated by a single annotator and independently corrected by another one. Inter-annotator agreement was ensured by frequent discussions joined by all annotators of the same category.

Annotation essentially followed the Dialogue Act Markup Language (DiAML) protocol (Bunt et al. 2010) with some additions for classes unavailable in it (including a range of categories based on unimodal observation). Multimodal pragmatic annotation included the following classes: communicative acts (constative, directive, com-missive, acknowledging, indirect), supporting acts (backchannel, politeness, repair), thematic control (topic initiation, topic elaboration, topic change), information (new information). Unimodal pragmatic annotation based on video alone included the classes of turn management (intend to start speaking, start speaking successfully, end speaking), attention (call, pay), agreement (default, full, partial, uncertainty, block, uninterested), deixis, information structure (received novelty). Unimodal annotation of physical events based on video included the classes of facial expressions (natural, happy, surprise, sad, recall, tense), gaze (directions of movement and their combi-nations, blink), eyebrows (directions), headshift (directions and their combinations), handshape (crossing fingers, open, flat, thumb-out etc.), touchmotion (tap, scratch, in combination with objects touched), posture (body, arms, head and their directions of movement), deixis (object, shape, self, addressee, measure), emotions (natural, happy, surprise, sad, recall, tense), emblem (attention, agree, doubt, disagree, refusal, block, doubt, more-or-less, etc.), other (for a gesture unclassified). Unimodal anno-tation of physical events based on audio included the classes of emotions (as above), discourse (turn management, backchannel), prosody (measured F0 values and their stylized rendering—rise, fall, stagnant, measured intensity values and their stylized rendering—increase, decrease, stagnant), overlapping speech, pause.

As for the HuComTech Corpus, it contains such a large amount of data, that the calculation of critical intervals for each pair of any two events and their combinations requires huge computational resources exceeding the capacity of everyday desktop computers (as of writing this paper, Theme is undergoing further development that will increase its computing capacity to the level that all data can be processed without any restrictions.) For now, we decided to select only a few classes of behavior gen-erally assumed to participate in evoking the observed events associated with uncer-tainty: the physical parameters of eyebrow, gaze, headshift, handshape and posture, as well as certain pragmatic contexts expected to accompany uncertainty: the multi-modal pragmatic classes of communicative acts, topic management, supporting act, information, the unimodal pragmatic classes of emblem, turn management, attention, agreement.

Uncertainty, the focus of our present work, is annotated as part of the unimodal annotation of the class of agreement, based on video only. It means that the annotators only followed what they saw on the video, without relying on any clue from the audio, either from the spoken text or its prosody. Even the speech tempo, the eventual pauses or brakes were only accessible indirectly, through the observed lip movements. The above listing of annotated classes shows that uncertainty was annotated as a label within the class of agreement. However, the annotators marked instances of observed uncertainty independently of any context either involving agreement or not, i.e. without expecting it to emerge as a sign of reference to a question, remark or backchannel that could eventually evoke uncertainty. It should be noted that the annotation only marked those instances which could be observed, and as such, they do not necessarily reflect the uncertainty as experienced by the speaker as their actual state of mind. Even with this obvious restriction, we believe that these data may also shed some light on uncertainty as experienced, even though there can be false positive and false negative value assignments as well. Our belief is based on the assumption that the interpretation of any state of mind, including that of uncertainty, is based on a context that is the product of a process of observation during which emerging events from a number of modalities are combined into a temporally developing dynamic pattern interpreted as representing the given state of mind. In this process the observer does not only consider temporal co-occurrences of various events, but calculates the statistical probability of the combination of a number of temporally adjacent or even non adjacent events to recognize them as a single pattern. This is a cognitive process that needs sufficient data and sufficient observation time to complete the calculation, and in addition, sufficient instances of one and the same pattern candidate to qualify for a particular pattern. In order to identify cues of uncertainty, we have used the software environment Theme (patternvision.com), specifically designed to discover such behavioral patterns. We will justify the use of Theme for our purpose in the next section.

6.3 Theme: a Tool for the Discovery of Hidden Patterns in Behavior

Theme is specifically designed to model the cognitive process of pattern discovery. It is based on the fundamental idea that any two events A, B can make a pattern if they occur within a critical interval satisfying a given probability, such as $p = 0.005$. (The critical interval for each pattern candidate is calculated using a cumulative binomial test; cf. Magnusson 2000.) A pattern in this sense does not require A and B to be adjacent, rather, it can allow for any number of other events to optionally occur between them. The concept of critical interval is especially important: it allows for A and B to occur at different time intervals every single time, but still form the same type of pattern—all occurrences falling within the same critical interval. Any pair of A, B not satisfying the requirement to fall within the given critical interval will not be considered an instance of the given pattern. Both the concept of critical interval and the optionality of adjacency are key to make Theme especially suitable for the recognition of patterns of behavior in everyday life: we recognize a series of events as a pattern even if its constituting events do not occur at exactly the same time (in fact they virtually never do)—instinctively we have a "sense" of this timing, more precisely, that of the critical interval within which A and B will be perceived as a single pattern. Also, the optionality of non-adjacency is equally important: it is hardly ever the case that no other event C comes between A and B, still, we consider C as "noise", as not belonging to the pattern A, B: if C does not satisfy the critical interval requirement, it does not make part of the pattern A, B. Our cognition does the job here: even though the presence of such an event C could be shown by some measurement, we do not even notice it, i.e. it is not perceived as part of the pattern A,B.

Uncertainty is such a state of mind whose context is especially variable: it varies depending on the actual social, psychological and informational state of the subjects, their goals and purposes, as well as the dynamically changing relation between them. It is then expected that uncertainty as a pattern can only be recognized across longer observation time, within statistically significant critical interval, and excluding the virtually ever present noise of insignificant events. The next sections will show how Theme discovers the behavioral patterns associated with uncertainty, often hidden from the naked eye. A last note before starting analyzing the actual data: whatever configuration of events Theme identifies as a pattern, it is the output of statistics, necessarily blind to what the data actually represent in real world. Even though we believe that Theme models the cognitive process of pattern discovery, the output of the model necessarily requires its proper review and evaluation based on our knowledge of this real world.

6.4   General Statistical Overview of the Corpus

The label for the event of uncertainty was annotated at least once in each of the 111 formal and informal recordings. However, due to the condition that a configuration of any number of events was only considered as a pattern if such a configuration (pattern candidate) occurred at least three times in a given recording, this condition was not met in several cases. As for the formal scenario, the number of files with no pattern of uncertainty was 34 (30.67% of all formal files), with males = 14 (12.6% of all subjects, 23.31% of males) and females = 20 (18.02% of all subjects, 39.22% of females). As for the informal scenario, the number of files with no pattern of uncertainty was 25 (22.52% of all informal files), with males = 13 (11.71% of all subjects, 21.64% of all males), females = 12 (10.81% of all subjects, 23.53% of all females). Accordingly, female subjects appeared to be less uncertain both in the formal and the informal conversation.

These data are interesting for two reasons. First, close to one-third of the formal recordings did not generate any pattern containing uncertainty, whereas in the formal recordings only just over one-fifth of them lacked such a pattern. This difference maybe due to the fact that the formal recordings tended to be relatively shorter than the informal ones, a difference that could eventually have an effect on the number of occurrences of any events, including uncertainty. However, it may have also been due to the difference in the respective scenarios: the formal recording was that of a job interview where the speaker (the subject applying for a job) usually tries to behave in a more formal way, more restrained, with fewer body movements and gestures, and with fewer overt expressions of the inner self.

Patterns are characterized by both level and length. As for level, it stands for the level of discovery they are identified at (level 1 = the initial level where a single item event A is associated with another single item event B to make the binary pattern A, B; level 2 = a pattern discovered at the next level, where the first half of the pattern is A, B, i.e. the pattern discovered at level 1, whereas the second half of the

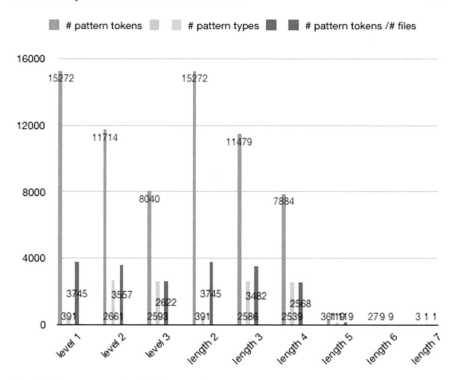

Fig. 6.1 T-patterns associated with uncertainty

pattern is either a single item event *C*, or another pair, also discovered at level 1; etc.). As for length, this value stands for the number of subpatterns a pattern consists of. Naturally, a level 1 pattern then has a length 2 with no subpatterns (since at level 1 a pair of single item events are associated into a pattern). At level 2 and further up, their length can vary depending on the complexity of at least one of the two halves of the given pattern allowing for the inclusion of at least a single event or a subpattern of a pair of single events; etc.

Figure 6.1 shows the corresponding data for all patterns associated with uncertainty in the corpus:

As expected, the most frequent patterns were discovered at the first level: 391 pattern types with 15272 occurrences, in all, 3745 samples of files. The latter figure means that a pattern occurred an average four times in one and the same dialogue. For these patterns, the maximum and minimum occurrences of pattern types and their standard deviations for each of the levels were found to be as follows: level 1: 29, 3, 1.16; level 2: 16, 3, 0.66; level 3: 7, 3, 0.28.

As for pattern length (the number of event types in one pattern), the longest pattern had a length of 7, i.e. it consisted of seven event types. Only one pattern of this length was found, in one dialogue only—in three samples, the minimum number to qualify for the status of a pattern. The number of patterns with a length of 6 was 9, occurring in 9 files, three times in each file. The 119 pattern types of length 5 also occurred

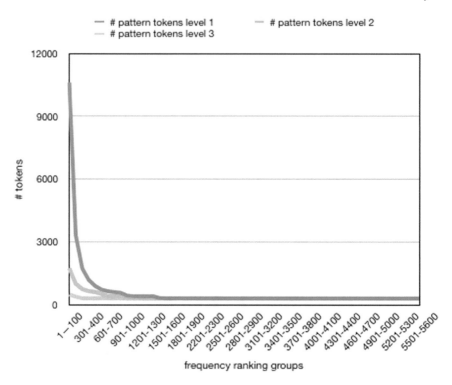

Fig. 6.2 Rank/frequency distribution of T-patterns related to uncertainty

only 3 times in a file for each pattern. The 2539 pattern types of length 4 occurred 7884 times, just over three (3.11) times in a file. Patterns with length 3 were already more frequent: the 2586 pattern types occurred 11479 times, an average 4.44 times in a dialogue. The most frequent patterns were those with length 2: the 2661 pattern types had a total frequency of 15272, occurring on average 39 times in a file.

Considering the above frequency data relating to either level or length, they suggest that their frequency decreases with the increase of structural complexity, a relation that might be somewhat similar to the rank/frequency distribution empirically formulated as Zip's Law (Zipf 1935, 1949). The first 10 most frequent pattern types (with a frequency between 558 and 150) constitute to just 0.18% of all pattern types, but they make up 7.17% of all tokens (2510 tokens). Almost half (48.52%) of all tokens (16994 occurrences) belong to just 7.37% of all pattern types (416 pattern types with frequencies between 558 and 10). Figure 6.2 shows this rank/frequency distribution of patterns related to uncertainty.

At the same time, as we see in Fig. 6.2, type-token frequency of patterns is also associated with complexity: a minimal pattern consisting of just two event types has a higher frequency (higher number of tokens) than a more complex pattern from a higher level (with more event types). However, if asked how many different files a given pattern type is found in, we find that it is not associated with token frequency; cf. Fig. 6.3:

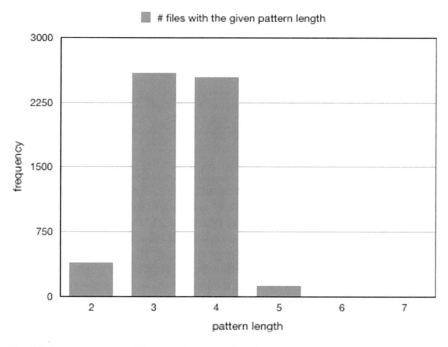

Fig. 6.3   Token frequency of T-patterns by pattern length

Fig. 6.4   Overall distribution
of pattern durations
associated with uncertainty

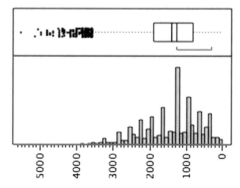

According to Fig. 6.3, patterns with just two event types are found in fewer files than those with 3 or 4 event types. However, patterns with more complexity (with 5 or more event types) occur in still fewer files. From this we suggest that the various behavioral patterns of uncertainty are mostly associated with a complex of 3 or 4 events rather than fewer or more. Considering that within Theme the critical interval for pattern search was set to a maximum of 1500 ms, it follows that the patterns of uncertainty discovered this way have a maximum duration of about 6 s—probably reasonable to our intuition. Let's have a look at the overall distribution of pattern durations associated with uncertainty; Fig. 6.4:

Table 6.1 Distribution of T-patterns and their durations by pattern length

	Length 2	Length 3	Length 4	Length 5	Length 6	Length 7
# Tokens	15272	11479	7884	361	27	3
Max duration	1481	2881	4321	5041	5521	3361
Upper quartile	1237	1921	2801	3521	3841	3361
Lower quartile	481	1240	1766	2321	2481	2641
Mean	812	1584	2268	2953	3225	2934
Stdev	393	554	676	839	966	378

The total number of patterns with uncertainty is 35026 pattern tokens. The maximum duration of a pattern with uncertainty is 5521 ms, the minimum is 6 ms. The upper quartile is 1921 ms, and the lower quartile is 837 ms, mean: 1417 ms, min: 6, stdev: 799 ms. The following table shows the distribution of patterns and their durations by pattern length (number of event types); cf. Table 6.1:

Pattern durations are not evenly distributed across and within length groups indicating that duration is affected by the particular event types the patterns are made of. Thus, even though we would expect patterns of length 7 to have a duration longer than any other, it turns out not to be necessarily so: in this particular case (just one pattern type in three occurrences, i.e. within one file only) uncertainty is associated with a series of very short head movements.

As mentioned earlier, the distribution of files with no patterns of uncertainty by gender of participants indicates that female subjects appeared in fewer files containing uncertainty, i.e. for all recordings, they showed fewer cases of uncertainty, than male subjects—both in the formal and the informal conversation paradigm. But, performing the one-way ANOVA t-test on files containing uncertainty we did not find a significant gender difference, regarding either the number of pattern types (formal: $p = 0.2607$, $DF = 88.808$; informal: $p = 0.1888$, $DF = 88.42$) or the number of pattern tokens (formal: $p = 0.2695$, $DF = 73.002$; informal: $p = 0.3305$, $DF = 80.105$). Also, no significant gender difference was found for pattern duration either (formal: $p = 0.6515$, $DF = 63.829$, informal: $p = 0.3335$, $DF = 75.119$). These data allow us to suggest that patterns of uncertainty must have some general characteristics, regardless of gender.

Naturally, the large number of pattern types (we identified 391 of them using *Theme*) are not equally found for all subjects or for similar frequencies. Out of the 222 files recorded with 111 subject for both the formal and informal conditions only 182 files contained events of uncertainty (either for the formal or the informal condition). It was a male subject, #119, who had the largest number of patter types, and, interestingly, all of them in the informal condition (i.e. he appeared to have a total lack of uncertainty in the formal job interview!). He was followed by a female subject, #072 with 641 pattern types, out of them 632 in the informal condition. The third most "uncertain" subject was #015, a male, with all the 478 pattern types in the job interview. With the variability that the above data show as an example of, it is not unexpected that the t-test shows no significance of gender difference in the ration between the number of pattern types in the formal and the informal condition ($p = 0.5596, DF = 68.63$).

Further looking at the data, out of the 111 subject only 69 were identified with events of uncertainty both in the formal and the informal condition. Interestingly, 51 (73.9%) of them were male, showing yet again that the male subjects showed more overall uncertainty in the conversations than the female ones. 58% of the 69 subjects (40 subjects) had a larger number of patterns with uncertainty in the informal condition than in the formal one (informal/formal, max: 101, min: 1.02 times more than in the formal condition, mean: 7.991, stdev: 18.90). 40.58% of them (28 subjects) had more patterns in the formal than in the informal condition (informal/formal, max: 0.83, min: 0.02, stdev: 0.27), and one subject had equal number of patterns in the two conditions.

After visiting further general characteristics of the corpus above, in the next section we will show the most frequent patterns expected to be characteristic of the perception of the state of mind by an observer.

6.5   Patterns of Uncertainty

Out of the 5258 pattern types of uncertainty the most frequent ones have a length of two, i.e. consist of two events, patterns identified at the first level of discovery. Out of them, 414 patterns start with uncertainty, and only 26 end with uncertainty. Table 6.2 lists patterns of uncertainty with the first event being uncertainty, limiting the list to those occurring in at least 30 files.

As can be seen in Fig. 6.5, the distribution of pattern durations is close to normal. However, those including gaze events tend to be both shorter and their values less variable than the ones without. This difference may be due to the very nature of annotation based on observation: an annotation of some event with a value requiring interpretation (such as uncertainty, elaboration, back channeling) is subject to a longer period of observation than the annotation of a visually unambiguous, rapid event, such as an eye, head or hand movement. It is, therefore, probably not by chance that certain unambiguous events of interaction, such as agreement, disagreement, refusal, calling attention etc. usually include the rapid movement of some body part (the head, eye or hand).

The role of the direction of gaze in association with uncertainty is clearly shown in the following patterns:

(1)   (up_agr,e,uncertainty v_gaze,b,forwards)
        (up_agr,b,uncertainty v_gaze,e,forwards)
        (up_agr,b,uncertainty v_gaze,b,down)
        (up_agr,e,uncertainty v_gaze,e,down)
        (up_agr,b,uncertainty v_gaze,b,up)

Table 6.2 T-patterns of uncertainty with the first event being uncertainty

Pattern type	# Files	# Tokens	Average # tokens/pattern	Average pattern duration (ms)
(up_agr,b,uncertainty up_agr,e,uncertainty)	100	558	5.58	1084.84
(up_agr,e,uncertainty v_gaze,b,forwards)	63	317	5.03	748.76
(up_agr,b,uncertainty v_gaze,e,forwards)	59	288	4.88	612.02
(up_agr,b,uncertainty mp_spcommact,b,constat)	54	249	4.61	752.81
(up_agr,b,uncertainty v_gaze,b,down)	52	234	4.5	639.49
(mp_spcommact,b,constat up_agr,e,uncertainty)	50	211	4.22	806.31
(up_agr,e,uncertainty mp_spcommact,e,constat)	42	173	4.12	1109.53
(mp_agcommact,e,directive up_agr,b,uncertainty)	40	150	3.75	598.27
(up_agr,e,uncertainty v_gaze,e,down)	36	169	4.69	673.66
(up_agr,e,uncertainty mp_agsuppact,b,backch)	33	143	4.33	884.97
(up_agr,b,uncertainty mp_agcommact,e,directive)	33	141	4.27	710.9
(mp_spcommact,e,constat up_agr,b,uncertainty)	33	129	3.91	789.84
(mp_spcommact,b,constat up_agr,b,uncertainty)	33	126	3.82	810.09
(mp_spsuppact,b,backch up_agr,e,uncertainty)	33	119	3.61	920.66
(up_agr,b,uncertainty mp_sptopic,b,t_elab)	32	144	4.5	959.3
(mp_spinf,b,new up_agr,e,uncertainty)	32	136	4.25	850.41
(up_agr,b,uncertainty v_gaze,b,up)	30	161	5.37	533.17
(up_agr,b,uncertainty mp_spinf,b,new)	30	138	4.6	872.88
(up_agr,b,uncertainty mp_sptopic,b,t_init)	30	128	4.27	718.5
(up_agr,e,uncertainty mp_sptopic,b,t_elab)	30	128	4.27	961.69
(up_agr,b,uncertainty mp_spcommact,e,constat)	30	119	3.97	926.04
(up_agr,b,uncertainty up_turn,b,startsp)	30	118	3.93	855.36

Fig. 6.5 Distribution of pattern durations associated with uncertainty

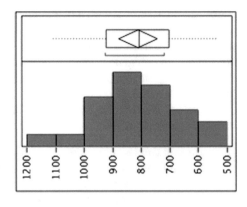

Namely, the speakers signaled the end of uncertainty ($up\_agr, e, uncertainty$) by beginning to look forward, i.e. to the agent ($v\_gaze, b, forwards$). As a variant, a pattern representing the end of uncertainty may also include the end of looking down

($v\_gaze, e, down$). In contrast, patterns representing the beginning of uncertainty either include starting to look down ($v\_gaze, b, down$) or up (v_gaze, b, up), or a preceding moment of either of the two latter events, starting to look away from the agent ($v\_gaze, e, forwards$). These patterns appear to clearly support one's generic behavioral stereotype about certainty/uncertainty: in a conversation, when we assert something, we usually look in the eyes of our counterpart, whereas in case of lack of certainty, we tend to look away. These examples also show that the visual expression of certainty/uncertainty does not exactly coincide with a given gaze event, instead, the latter somewhat follows uncertainty, even if just within a fraction of a second. From this it also follows that the event of looking down or up or simply away from the partner does not by itself lend the sense of uncertainty, even if stereotypically it might be associated with it. Instead, the interpretation of the given state of mind must then be based on additional visual signals of behavior. In turn, the interval (however short it is) between the interpretation of uncertainty and the perception of the change of gaze direction might rather reflect the duration of a cognitive process associated with uncertainty but reaching beyond its perception—as technically supported by the fact that the annotation of uncertainty was done independently from that of gaze direction.

Let us have a closer look at some of the actual contexts the above patterns are found in the corpus (in most cases we'll refer to patterns by the speaker; the agent's utterance is only given when it has a role in the context of the particular behavior; each file is identified by a number with the prefix "f" for formal, and "i" for informal conversation, followed by the location of the pattern in ms):

(2) f003: 13790-17141
(up_agr,e,uncertainty v_gaze,b,forwards)
(agent) b meséljen az előző munkahelyeiről! (Speak about your previous work!)
(speaker) p hát munkahelyem még nem volt. (Well, I have not worked before.)

(3) f007: 23145-24918
(up_agr,b,uncertainty v_gaze,e,forwards)
(agent) akkor a tanulmányairól. b (then törölni about your studies.)

(4) f017: 45681-48878
(up_agr,b,uncertainty v_gaze,b,down)
(speaker) %s p voltam %s p egy alkalommal %o telefonos operátor (I was once a telephone operator)

(5) f057; 59418-63335
(up_agr,e,uncertainty v_gaze,e,down)
(speaker) igazából b %o müvészeti képzésben vettem részt. (in fact I took part in art education)

(6) f114; 59666-61426
(up_agr,b,uncertainty v_gaze,b,up)
(agent) (or do you plan to extend your studies to other languages?)
(speaker) %o p igen, tervezem (yes, I plan)

Table 6.3 T-patterns with uncertainty as their second event

Pattern type	# Files	# Tokens	Average # tokens/pattern	Average pattern duration (ms)
(v_gaze,e,forwards up_agr,b,uncertainty)	28	112	4	713.86
(v_gaze,e,up up_agr,e,uncertainty)	28	112	4	4
(v_gaze,b,down up_agr,b,uncertainty)	27	105	3.89	577
(v_gaze,e,down up_agr,e,uncertainty)	26	110	4.23	662.82
(v_gaze,b,forwards up_agr,e,uncertainty)	25	85	3.4	593
(v_gaze,b,up up_agr,e,uncertainty)	25	104	4.16	747.92
(v_gaze,e,forwards up_agr,e,uncertainty)	25	93	3.72	3.72
(v_gaze,b,down up_agr,e,uncertainty)	24	111	4.63	940.82
(v_gaze,b,forwards up_agr,b,uncertainty)	20	68	3.4	838.35
(v_gaze,b,up up_agr,b,uncertainty)	15	58	3.87	541.69
(v_gaze,e,down up_agr,b,uncertainty)	13	48	3.69	912.67
(v_gaze,b,left up_agr,b,uncertainty)	8	26	3.25	616.38
(v_gaze,e,right up_agr,e,uncertainty)	7	26	3.71	3.71
(v_gaze,b,blink up_agr,b,uncertainty)	6	32	5.33	716
(v_gaze,b,right up_agr,e,uncertainty)	6	27	4.5	821.74
(v_gaze,b,right up_agr,b,uncertainty)	5	17	3.4	448.06
(v_gaze,e,left up_agr,b,uncertainty)	5	18	3.6	3.6
(v_gaze,b,blink up_agr,e,uncertainty)	4	15	3.75	971.67
(v_gaze,e,left up_agr,e,uncertainty)	4	13	3.25	3.25
(v_gaze,e,right up_agr,b,uncertainty)	4	13	3.25	3.25
(v_gaze,b,left up_agr,e,uncertainty)	3	10	3.33	609
(v_gaze,e,blink up_agr,e,uncertainty)	3	13	4.33	905.62
(v_gaze,e,blink up_agr,b,uncertainty)	2	9	4.5	267.67
(v_gaze,e,up up_agr,b,uncertainty)	2	6	3	3

The 26, 2-element patterns with uncertainty as their second event demonstrate the specific role of gaze in the perception of uncertainty; cf. Table 6.3 (the complete list is shown due to the restricted number of occurrences):

The first we can notice is the very short duration (4 ms or under) of 7 patterns, all starting with the end of some gaze movement:

(7)   (v_gaze,e,up up_agr,b,uncertainty)
       (v_gaze,e,left up_agr,e,uncertainty)
       (v_gaze,e,right up_agr,b,uncertainty)
       (v_gaze,e,left up_agr,b,uncertainty)
       (v_gaze,e,right up_agr,e,uncertainty)
       (v_gaze,e,forwards up_agr,e,uncertainty)
       (v_gaze,e,up up_agr,e,uncertainty)

Due to these very short durations we assume that the change of the direction of the gaze in these cases virtually happens at the same time as the beginning or end of the moment uncertainty is visually observed.

It is interesting to observe that in some patterns the very same gaze is associated with opposite values of uncertainty (begin [b] or end [e]):

(8)   (v_gaze,e,forwards up_agr,b,uncertainty)
      (v_gaze,e,up up_agr,e,uncertainty)

      (v_gaze,b,forwards up_agr,e,uncertainty)
      (v_gaze,e,forwards up_agr,e,uncertainty)

      (v_gaze,b,up up_agr,b,uncertainty)
      (v_gaze,b,up up_agr,e,uncertainty)

      (v_gaze,b,down up_agr,b,uncertainty)
      (v_gaze,e,down up_agr,b,uncertainty)
      (v_gaze,b,down up_agr,e,uncertainty)

      (v_gaze,b,right up_agr,e,uncertainty)
      (v_gaze,e,right up_agr,e,uncertainty)

These data suggest that a given gaze movement is not directly associated with the subsequent observed uncertainty. However, the fact that gaze and uncertainty still constitute to statistically significant patterns, points to the existence of a more complex cognitive process here: uncertainty is being built up so that the change in the movement of gaze reflects the end of its initial (or at least earlier) phase, and uncertainty itself represents its final phase.

As for blinking, another item annotated with regard to gaze, whereas it was not observed as a second member of patterns of uncertainty, it does appear as the initial part of a pattern, allowing us to attribute to it some function. Indeed, even if—similarly to other gaze movements—it can also be found with opposite values of uncertainty, such as

(9)   (v_gaze,b,blink up_agr,b,uncertainty)
      (v_gaze,b,blink up_agr,e,uncertainty)
      (v_gaze,e,blink up_agr,e,uncertainty)
      (v_gaze,e,blink up_agr,b,uncertainty)

blinking may again represent the end of an initial part of a more complex cognitive process leading up to uncertainty. Its lack as the final part of a pattern, on the other hand, suggests that, at least, it is not a direct follow-up of uncertainty.

Next, let us examine patterns of uncertainty at level 2, patterns consisting of three events. Out of a total of 2587 patterns 2341 have uncertainty as their first, 445 as their second (middle), and 362 as their second (last) event. (NB! The sum of the number

Table 6.4 Most frequent three-event patterns with uncertainty as their first event

Pattern type	# Files	# Tokens	Average # tokens/ pattern	Average pattern duration (ms)
((up_agr,b,uncertainty up_agr,e,uncertainty) v_gaze,b,forwards)	11	43	3.91	1974.95
((up_agr,b,uncertainty mp_spsuppact,b,backch) mp_spsuppact,e,backch)	10	40	4	1713
(up_agr,b,uncertainty (v_gaze,b,up v_gaze,e,up))	10	37	3.7	1322.08
((up_agr,b,uncertainty mp_sptopic,b,t_init) mp_sptopic,e,t_init)	9	32	3.56	1671
((up_agr,e,uncertainty mp_agsuppact,b,backch) mp_agsuppact,e,backch)	8	26	3.25	1944.15
((up_agr,b,uncertainty v_gaze,e,forwards) up_agr,e,uncertainty)	7	25	3.57	1370.6
((up_agr,b,uncertainty mp_spinf,b,new) up_agr,e,uncertainty)	7	23	3.29	1475.78
((up_agr,e,uncertainty mp_spsuppact,b,backch) mp_spsuppact,e,backch)	7	24	3.43	1561
(up_agr,b,uncertainty (up_agr,e,uncertainty mp_sptopic,b,t_elab))	6	20	3.33	1441
(up_agr,b,uncertainty (up_agr,e,uncertainty up_att,b,calling))	6	20	3.33	1825
(up_agr,b,uncertainty (up_agr,e,uncertainty v_gaze,b,forwards))	6	19	3.17	1836.79
(up_agr,b,uncertainty (mp_spinf,b,new up_agr,e,uncertainty))	6	19	3.17	1567.32
((up_agr,b,uncertainty mp_sptopic,b,t_init) up_agr,e,uncertainty)	6	20	3.33	1249
((up_agr,e,uncertainty up_att,b,calling) up_att,e,calling)	6	20	3.33	1697
((up_agr,b,uncertainty mp_agsuppact,e,backch) up_agr,e,uncertainty)	6	20	3.33	1409
(up_agr,e,uncertainty (v_gaze,b,forwards v_gaze,e,forwards))	6	22	3.67	2059.18
(up_agr,b,uncertainty (up_agr,e,uncertainty mp_sptopic,e,t_init))	5	16	3.2	1721
((up_agr,b,uncertainty mp_agcommact,e,directive) v_gaze,b,down)	5	17	3.4	1342.18
((up_agr,b,uncertainty up_agr,e,uncertainty) mp_sptopic,b,t_elab)	5	18	3.6	1707.67
((up_agr,e,uncertainty mp_sptopic,b,t_elab) v_gaze,b,forwards)	5	19	3.8	1546.23
(up_agr,b,uncertainty (v_gaze,e,blink v_gaze,e,down))	5	16	3.2	2191
(up_agr,b,uncertainty (mp_agcommact,e,directive mp_spcommact,b,constat))	5	17	3.4	753.94
((up_agr,b,uncertainty mp_agcommact,e,directive) v_gaze,e,forwards)	5	17	3.4	956.29
(up_agr,b,uncertainty (mp_spcommact,b,constat up_agr,e,uncertainty))	5	15	3	1366.33

of patterns with the three different positions of uncertainty (3148) exceeds the total number of patterns (2587)—due to the fact that many patterns contain uncertainty both as the first and the second, or the second and the third event, the beginning of uncertainty preceding the end of uncertainty, respectively, and so one and the same pattern is considered in two classes, instead of just one.

Table 6.4 shows the most frequent three-event patterns with uncertainty as their first event, occurring in at least 5 files:

Having a closer look at the table we find that out of the 24 patterns 20 start with the beginning of uncertainty. Among them in the most frequent three uncertainty is followed by the rapid events of gaze, or backchannel by the same actor:

(10)   ((up_agr,b,uncertainty up_agr,e,uncertainty) v_gaze,b,forwards)
        ((up_agr,b,uncertainty mp_spsuppact,b,backch) mp_spsuppact,e,backch)
        (up_agr,b,uncertainty (v_gaze,b,up v_gaze,e,up))

The above patterns are found in such contexts as:

(11)   i031, 16753-17983

((up_agr,b,uncertainty up_agr,e,uncertainty) v_gaze,b,forwards)

(speaker) Jó, igen, nyilván (Yes, of course, naturally)

(12)   i119, 75528-76099

((up_agr,b,uncertainty mp_spsuppact,b,backch) mp_spsuppact,e, backch)

(speaker) %s mégis (still)

Backchannel by the agent is also among the frequent patterns—as expected, following the end of the speaker's uncertainty:

(13)   f036, 57789-59856

((up_agr,e,uncertainty mp_spsuppact,b,backch) mp_spsuppact,e, backch)

(speaker) Nagyon szeretem ezt az emberi erőforrást (I like this human resource very much)

When the speaker stops being uncertain, however, other patterns include further behavior by the same speaker: looking forward or calling the attention of the agent (these two are probably related: looking forward may be an introductory event to calling attention):

(14)   f071, 531455-535829

(up_agr,e,uncertainty (v_gaze,b,forwards v_gaze,e,forwards))

(speaker) p erre is van, %s úgymond egy p igény az embernek (in fact one also has the desire for it)

(15)   i12, 34817-36140

((up_agr,e,uncertainty up_att,b,calling) up_att,e,calling) (speaker) meg nem is tudom, így (and I don't know, so)

Of course, uncertainty may be a reflection to a variety of contexts. It may precede the introduction of a new topic:

(16)   f073, 32276-36887

((up_agr,b,uncertainty mp_sptopic,b,t_init) up_agr,e,uncertainty)

(speaker) ott %o főleg ilyen gyógyszerválogatá%sról szólt a munka (there the task was to select drugs)

or it can also be the response to an utterance by the other actor. The following pattern is an example for a mini-situation: the speaker reacts with uncertainty to the agent starting giving some feedback to what the speaker has just said (backchannel), but this uncertainty is short lived: it ends with the end of the speaker's feedback:

(17)   f018, 53139-57571
         ((up_agr,b,uncertainty mp_agsuppact,e,backch) up_agr,e,uncertainty)
         (speaker) és %o hát elég sok területet érintettünk, így %o %s (we covered
         quite a few topics)
         (agent) %s

The following two examples show how the speaker prepares for elaborating a new
topic in the discourse. In both cases the speaker starts elaborating after some hesita-
tion (after overcoming uncertainty), with the second pattern also including the gaze
directed towards the agent:

(18)   i110, 56640-562240
         (up_agr,b,uncertainty (up_agr,e,uncertainty mp_sptopic,b,t_elab))
         (speaker) szokott lenni, csak %o nem szoktam megjegyezni I b (it occurs but
         I do not usually remember)

(19)   f126, 112960-114800
         ((up_agr,e,uncertainty mp_sptopic,b,t_elab) v_gaze,b,forwards)
         (speaker) hát a munkától függ (well, it depends on the job)

Patterns including a directive by the agent and response by the speaker also capture
how we intuitively imagine such a situation. In the next three patterns the speaker
responds to the directive of the agent (the beginning of the directive falls outside
the pattern due to the varied length of the directive itself): in the first pattern the
speaker ends looking forward (while they were listening to the directive, they were
looking towards the agent, but now, starting looking elsewhere they were preparing
for a response); the second pattern captures just this next moment: looking down –
preparing for the response; whereas the last pattern captures the response itself: the
speaker starts a communicative act:

(20)   f114, 101440-102400
         ((up_agr,b,uncertainty mp_agcommact,e,directive) v_gaze,e,forwards)
         (agent) Valamilyen szakmai gyakorlata van-e? (I wonder if you have any job
         experience?)

(21)   f123, 621440-622800
         ((up_agr,b,uncertainty mp_agcommact,e,directive) v_gaze,b,down)
         (agent) és melyik ami a személyiségéhez közelebb áll? (and which is closer
         to your personality?)
         (speaker) talán az e– az egyed (perhaps the in– the individual)

(22)   f067, 85440-86720
         **(up_agr,b,uncertainty (mp_agcommact,e,directive mp_spcommact,b, con-
         stat))**
         (agent) *tát mi az a munkakör, amit távol állónak érez? (so, which is the job
         you feel far from you?)

Table 6.5 T-patterns in which uncertainty is the middle constituent in three-part patterns

Pattern type	# Files	# Tokens	Average # tokens/ pattern	Average pattern duration (ms)
((mp_spcommact,b,constat up_agr,e,uncertainty) v_gaze,b,forwards)	12	36	3	1578.67
((up_agr,b,uncertainty up_agr,e,uncertainty) v_gaze,b,forwards)	11	33	3	1974.95
((mp_spcommact,e,constat up_agr,b,uncertainty) up_agr,e,uncertainty)	6	18	3	1905
((mp_sptopic,b,t_elab up_agr,b,uncertainty) up_agr,e,uncertainty)	5	15	3	1905.76
((mp_agsuppact,e,backch up_agr,b,uncertainty) up_agr,e,uncertainty)	5	15	3	1743.18
((mp_agcommact,e,constat up_agr,b,uncertainty) v_gaze,b,down)	5	15	3	1767.12
((up_agr,b,uncertainty up_agr,e,uncertainty) mp_sptopic,b,t_elab)	5	15	3	1707.67
((mp_agcommact,e,directive up_agr,b,uncertainty) v_gaze,b,down)	5	15	3	934.06
((up_att,e,paying up_agr,b,uncertainty) v_gaze,e,forwards)	5	15	3	1111
((mp_agsuppact,b,backch up_agr,e,uncertainty) mp_agsuppact,e,backch)	5	15	3	1321

Next, let us have a look at patterns in which uncertainty is the middle constituent in three-part patterns, occurring at least in 5 files (see Table 6.5).

These patterns capture essentially the same contexts we saw above uncertainty appears in: those involving constative communicative acts, backchannel, attention, topic elaboration—events strongly associated with an exchange of turns. The two most frequent patterns include forward gaze movement, the strongest indication of engagement in a face-to-face interaction:

(23)   f039, 211280-213120

((mp_spcommact,b,constat up_agr,e,uncertainty) v_gaze,b,forwards)
(speaker) meg %o *má nem szívesen dolgoznék %o (and I would not be happy to work)

(24)   i020, 795276-796716

((up_agr,b,uncertainty up_agr,e,uncertainty) v_gaze,b,forwards)
(speaker) vagy arra buzdítják az embereket, hogy másoknak ártsanak (or some inspire people to hurt others)

Uncertainty also develops as a reaction to the agent's backchannel or other communicative act. In all these cases the speaker is waiting until the agent's action comes to an end:

(25)   f071, 587089-588160

((mp_agsuppact,e,backch up_agr,b,uncertainty) up_agr,e,uncertainty)
(agent) uhum

(26)   i076, 402712-404800

((mp_agcommact,e,constat up_agr,b,uncertainty) v_gaze,b,down)
(agent) akkor már kevesebb ideje marad b s%ajnos. (then he will have less time left, unfortunately)

Table 6.6 T-patterns in which uncertainty is the final constituent in three-part patterns

Pattern type	# Files	# Tokens	Average # tokens/ pattern	Average pattern duration (ms)
((v_gaze,b,forwards v_gaze,e,forwards) up_agr,b,uncertainty)	7	21	3	1694.79
((up_agr,b,uncertainty v_gaze,e,forwards) up_agr,e,uncertainty)	7	21	3	1370.6
((up_agr,b,uncertainty mp_spinf,b,new) up_agr,e,uncertainty)	7	21	3	1475.78
((v_gaze,b,up v_gaze,e,up) up_agr,e,uncertainty)	6	18	3	1624.33
((mp_spcommact,e,constat up_agr,b,uncertainty) up_agr,e,uncertainty)	6	18	3	1905
((up_agr,b,uncertainty mp_sptopic,b,t_init) up_agr,e,uncertainty)	6	18	3	1249
((up_agr,b,uncertainty mp_agsuppact,e,backch) up_agr,e,uncertainty)	6	18	3	1409
((mp_sptopic,b,t_elab up_agr,b,uncertainty) up_agr,e,uncertainty)	5	15	3	1905.76
((mp_agsuppact,e,backch up_agr,b,uncertainty) up_agr,e,uncertainty)	5	15	3	1743.17
((up_agr,b,uncertainty up_turn,b,startsp) up_agr,e,uncertainty)	4	12	3	1578.14
((up_agr,b,uncertainty mp_sptopic,e,t_elab) up_agr,e,uncertainty)	4	12	3	1305.62
((up_agr,b,uncertainty mp_spcommact,b,constat) up_agr,e,uncertainty)	4	12	3	1404.08
((up_agr,b,uncertainty mp_sptopic,b,t_elab) up_agr,e,uncertainty)	4	12	3	1494.33
((up_agr,b,uncertainty mp_spsuppact,e,backch) up_agr,e,uncertainty)	4	12	3	1361
((mp_sptopic,b,t_init up_agr,b,uncertainty) up_agr,e,uncertainty)	4	12	3	1943.86
((up_att,e,calling up_agr,b,uncertainty) up_agr,e,uncertainty)	4	12	3	1947.66
((v_gaze,b,forwards v_gaze,b,down) up_agr,b,uncertainty)	4	12	3	1886
((up_agr,b,uncertainty v_gaze,b,up) up_agr,e,uncertainty)	4	12	3	1537

(27)   i126, 297189-298400
((mp_agcommact,e,directive up_agr,b,uncertainty) v_gaze,b,down)
(speaker) fú ott meg már rég– még régebben voltam. | (well there I haven't
been lately)

The next pattern, however, reflects a situation where the speaker stops being uncertain
as soon as the agent starts giving a supportive backchannel, still before finishing it:

(28)   i031, 38916-39556
((mp_agsuppact,b,backch up_agr,e,uncertainty) mp_agsuppact,e,backch)
(agent) uhum

Finally, let us look at patterns in which uncertainty is the final constituent in three-part
patterns, occurring at least in 4 files (see Table 6.6).

Two of the most frequent patterns are associated with gaze. That looking away
from the agent does not necessarily mean uncertainty is shown in these patterns:

(29)   f048, 292000-293760
((v_gaze,b,forwards v_gaze,e,forwards) up_agr,b,uncertainty)
(speaker) [%o semmiképpen nem tudnék irodában dolgozni, monitor előtt]
emberektől elszigetelve ([by no means could I work in an office, in front of
a monitor] isolated from people)

(30)   f114, 115520-117120
        ((up_agr,b,uncertainty v_gaze,e,forwards) up_agr,e,uncertainty)
        (speaker) de konkrétan állásom még így, hogy szerződött, nem volt (but
        concretely, so far I have not had a contracted job)

Accordingly, in the first example the speaker first looks forwards then away, and the
behavior will evolve into uncertainty; in the second, the beginning of uncertainty
is associated with the gaze moving away from the agent, but then the speaker stops
being uncertain. It indicates that gaze direction by itself may have some stereotypical
interpretation, but—as in virtually all other behaviors—the interpretation of an actual
situation should rather be based on the identification of multimodal patterns. The
multiple movement of gaze, however, may suggest uncertainty, also following our
intuition as well. As a condition, looking forwards and/or down should most probably
be relatively long; the following pattern is among the longest recorded above (almost
2 s):

(31)   i020, 610316-611916
        ((v_gaze,b,forwards v_gaze,b,down) up_agr,b,uncertainty)
        (speaker) csak kiesett egy pár dolog (only a few things fell out)

However, the single but relatively long (here: 1624.33 ms) gaze direction (looking up)
may be specific to uncertainty. Looking at the following pattern we may intuitively
associate it with a situation where the speaker is preparing for an answer (and is
so far uncertain about it), reflected by his/her upward gaze direction; with the gaze
changing this direction the response is here (and the uncertainty disappears):

(32)   i006, 399175-400935
        ((v_gaze,b,up v_gaze,e,up) up_agr,e,uncertainty)
        (speaker) és b ez nagyon rossz érzés volt (and it was a very bad feeling)

One more example of ending uncertainty: when, after some deliberation, the speaker
starts speaking, his/her uncertainty may also disappear, as shown here:

(33)   f013, 302699-304619
        ((up_agr,b,uncertainty up_turn,b,startsp) up_agr,e,uncertainty)
        (speaker) hát b attól – a munkától függ igazából (well, it really depends on
        the job)

6.6   Summary

Face-to-face communication is essentially multimodal: it involves a complex of verbal and non-verbal means of interaction to convey both the delivery and the reception
of information. This multimodal complex of resources of information is the basis for
expressing our thoughts, intentions and sentiments. Even though the verbal modality
with its lexicon, syntax and semantics is a highly powerful part of this complex, in

order to actually understand the larger context of a given conversation one needs to go beyond it and consider all pieces of information available from any other modality. The temporal dynamics of an interaction requires to process information as a function of time: an event that happens at a given time may be related to another event at a different time, preceding or following it. This relation can be captured as a pattern, which, in turn, is a minimal functional building block of the whole of a given interaction. Accordingly, singular events, such as the word "yes", or a blink of an eye is not by itself meaningful, it receives its "meaning", i.e. communicative function only as part of a pattern.

When studying human behavior using data from the multimodal HuComTech Corpus, our aim was to discover multimodal patterns of behavior that may be characteristic of certain key features of a conversation. One of these features is associated with how certain the participants of a given dialogue appear to be when giving or receiving information. Our present research was aimed at the discovery of behavioral patterns associated with uncertainty. For pattern discovery Theme, a computer framework, specifically designed for the study of communication was used. Due to technical constraints we restricted our search for patterns with a maximum distance of 1500 ms between two events as candidates for a minimal pattern. On the basis of data from a large corpus (111 subjects, 2 communicative conditions, 50 hours of recordings) we found that there was no significant gender difference between subjects regarding the number of pattern types or the number of pattern tokens, i.e. the actual occurrences of patterns. A number of files did not show patterns of uncertainty. It was found that the more complex a pattern is, the fewer its occurrences. However, it turned out that male subjects showed uncertainty in more files (recordings) than women, both in the formal and the informal communicative condition. Two lists of patterns of uncertainty identified in the corpus were presented: patterns with two events, and patterns with three events, each of them further specified as to the position of uncertainty in the pattern. It was found that uncertainty formed a pattern with other interpretive events more frequently than with visually identifiable events, such as gaze. Probably due to the predefined short, 1500 ms critical interval between two events, patterns mostly included events associated with just one actor, however interesting patterns reflecting short interactions between the speaker and the agent were also identified. With the further development of our hardware and software environment we expect to identify more complex patterns of behavior, ranging over a longer span of time with more hierarchical dependencies.

References

Bach K, Harnish R (1979) Linguistic communication and speech acts. MIT Press, Cambridge, Mass
Bunt H, Alexandersson J, Carletta J, Choe J, Fang AC, Hasida K, Lee K, Petukhova V, Popescu-Belis A, Romary L, Soria C, Traum D (2010) Towards an ISO standard for dialogue act annotation. In: Proceedings of the seventh conference on international language resources and evaluation (LREC10, Valletta)

Hunyadi L (2011) A multimodális ember-gép kommunikáció technológiái – elméleti modellezés és alkalmazás a beszédfeldolgozásban [technologies of multimodal human-machine communication—theoretical modeling and application in speech processing]. In: Enikő Németh T (ed) Ember-gép kapcsolat. A multimodális ember-gép kommunikáció modellezésének alapjai [The relationship between human and machine. The bases of the modelling of multimodal communication between human and machine]. Tinta Könyvkiadó, Budapest, pp 15–41

Hunyadi L, Földesi A, Szekrényes I, Staudt A, Kiss H, Abuczki A, Bódog A (2012) Az ember-gép kommunikáció elméleti-technológiai modellje és nyelvtechnológiai vonatkozásai. [a theoretical-technological model of human-machine communication and its reference to language technology]. In: Kenesei I, Prószéky G, Váradi T (eds) Általános Nyelvészeti Tanulmányok XXIV. Akadémiai, Budapest

Magnusson M (2000) Discovering hidden time patterns in behavior: T-patterns and their detection. Behav Res Methods Instrum Comput 32(1):93–110

Magnusson MS, Burgoon J, Casarrubea M (eds) (2016) Discovering hidden temporal patterns in behavior and interaction: T-pattern detection and analysis with THEME. Springer, New York

Németh TE (ed) (2011) Ember-gép kapcsolat. A multimodális ember-gép kommunikáció modellezésének alapjai [The relationship between human and machine. The bases of the modelling of multimodal communication between human and machine]. Tinta Könyvkiadó, Budapest

Pápay K, Szeghalmy S, Szekrényes I (2011) Hucomtech multimodal corpus annotation. Argumentum 7:330–347

Sperber D, Wilson D (1986) Relevance: communication and cognition. Blackwell, Oxford

Tversky A, Kahneman D (1973) Judgement under uncertainty: heuristics and biases. ONR technical report. National technical information service

Wilson D, Sperber D (2002) Relevance theory. In: Horn LR, Ward G (eds) The handbook of pragmatics. Blackwell, Oxford, pp 607–632

Wilson D, Sperber D (2012) Meaning and relevance. Cambridge University Press, Cambridge

Zipf GK (1935) The psychobiology of language. Houghton-Mifflin, New York

Zipf GK (1949) Human behavior and the principle of least effort. Addison-Wesley, Reading MA (USA)

Chapter 7
Watching People Talk; How Machines Can Know We Understand Them—A Study of Engagement in a Conversational Corpus

Nick Campbell

Abstract This paper describes an examination of the HuComTech Corpus of multimodal interactions from the viewpoint of an automated dialogue system. Specifically, it looks at the difference between formal and informal interactions from the point of view of image processing. We show that an autonomous dialogue system is able to make inferences about the engagement states of its interlocutor in order to function efficiently without complete dialogue understanding. The paper shows that a very simple image processing algorithm is capable of distinguishing engagement states with a high degree of reliability. We can infer from the results that future autonomous agents might use a similar method to estimate the degree of engagement a human interlocutor has in their conversations.

7.1 Introduction

While the ubiquitous use of the telephone confirms that humans can easily communicate through voice without seeing each other, many people, especially non-native speakers, prefer to see the person they are talking with, in order to verify the efficient flow of communication. This is perhaps especially relevant for spoken dialogue systems used in 'caring' situations; not the task-based Call-Centre type of interaction, but the home or hospital situation where people are interacting regularly with a virtual agent and will engage in non-task-based modes of conversation quite frequently.

These 'social' dialogue systems face a particularly difficult problem where often the words of a conversation are insufficient for a full understanding of the intentions of the speaker. In these cases, a measure of 'engagement' can be helpful in estimating how best to proceed and what to say next. This paper will explore a way of using the

N. Campbell (✉)
Speech Communication Lab, Trinity College Dublin, Dublin 2, Ireland
e-mail: nick@tcd.ie

© Springer Nature Switzerland AG 2020
L. Hunyadi and I. Szekrényes (eds.), *The Temporal Structure of Multimodal Communication*, Intelligent Systems Reference Library 164,
https://doi.org/10.1007/978-3-030-22895-8_7

eyes (or image processing) to make an intelligent guess at how involved or engaged a partner might be. We use the HuComTech Corpus (Hunyadi et al. 2016) of dialogue recordings as our material and OpenCV (Bradski 2000) (free software from Intel) for the image processing.

7.2 The HuComTech Corpus

As described in the MetaShare site[1] that hosts it, the HuComTech multimodal corpus "consists of about 50 h of video and audio recordings of 111 formal dialogues (simulated job interviews) and 111 informal but guided dialogues. The language of the recordings is Hungarian. The participants were university students aged 19–27, female 54 and male 67. The corpus was annotated for video (facial expressions, instances of eyebrows, gaze, headshift. handshape, touchmotion and posture) and audio (emotions, discourse, prosody and textual transcriptions). Its unique features in a wider comparison include its special attention to pragmatics focusing on a comparative study of the unimodal versus multimodal features of communication (as compared to multimodality alone) as well as the study of the syntax and prosody of spoken language with respect to the above wide range of multimodal characteristics. The data can be queried in ELAN and in our web-based SQL database". Full details of the corpus can be found in the working paper (Pápay et al. 2011).

The corpus was originally designed for the study of prosody. Szekrényes (2014) reports on the differences of measured and perceived aspects of prosodic data and a prosodic annotation method which is implemented in the HuComTech corpus based on the results of existing applications and the psychoacoustic model of tonal perception : "Prosody is an important component of multimodal communication. Appropriate interpretation of speech would be less effective and sometimes impossible without the information expressed by prosodic features. Complementing other sources of information, computer recognition of various communicative events and individual characteristics of the speaker are also based on prosody where measured physical features (such like fundamental frequency and intensity) can be used as parameters for further operations".

The present work complements the study of prosody by positing a form of 'visual prosody' in which physical body dynamics complement acoustic vocal dynamics in a similar manner. We examine not the tone-of-voice, but the flow of the hands and the movements of the head and body with the intention of 'reading beyond the words' and of exploring the kinds of meaning that a multimedia corpus can convey.

---

[1] http://metashare.nytud.hu/.

7.3 The Role of the Facilitator

In the original design of the corpus it was not envisaged (or so I believe) that the facilitator should also become a subject of analysis with respect to the resulting data. A staff member from the laboratory operated the recorders, signalled the start and end of each conversation and elicited a series of responses from the subjects in either the formal or the informal situation. The separate task of reading 15 sentences aloud did not involve such two-way interaction.

However, the facilitator is the one factor that is common to ALL conversations. She had to interact with all participants in a minimal and unobtrusive way but to ensure that the conversations flowed smoothly and that the subjects 'performed' as required. She was there to do a job; but she is also a human being and reacts to her subjects in a human way, laughing if they joke, and listening quietly if they are speaking. Similar to a TV news facilitator in street interviews, her role was to make the people speak without intruding herself as a participant. She was very successful at remaining unobtrusive. And she was present in every recording, with one camera directed at her throughout.

7.4 Analysis of the Engagement

There were three video cameras in the room where the interviews took place. Two filming the subject (left and right, behind the facilitator) and one to the rear right, behind the subject, recording the facilitator herself. It is this third camera that we used for the analysis. We can see the behaviour of the facilitator from before the start to after the end of each conversation. She makes much movement as they enter and leave but sits very still in her seat as they speak. She is of course listening. And perhaps 'responding' to their speech, but in such a way as to minimise her influence on the behaviour of the subjects.

We first proposed (as logic and previous experience suggested Haider et al. 2016) to track her face and head movements to measure her engagement in the conversations. OpenCV has an excellent face-detection programme in its sample directory that has been used before in many of our experiments (Campbell 2009). However, on closer observation of her movements during the dialogues, we noticed that it is the hands rather than the face that carry much of the important information in this situation. As she sits listening to (monitoring?) the speech of her subjects, she twiddles her fingers or sits impassively; moving her hands expressively whenever she speaks back to them. Kendon has written extensively on the use of gesture in conversational interaction (Kendon 1990), but this is the 'silent' movement of an observer, rather than the actions of a participant. It tells us much about her cognitive states as she monitors their performance.

In order to quantify this movement, we used the Lucas–Kanade tracker from OpenCV (version 3.3) (Bradski 2000) to process the video images from Camera

Fig. 7.1 The third camera's view of the facilitator as imaged by OpenCV 3.3, using the default settings of the Python lk_track.py software for movement detection and tracking. The green dots are assigned automatically by the programme as 'points_good_to_track', and the small lines attached to some of the points show direction of recent movement. Dots without lines are unchanged from the previous frame

3 (facing the facilitator, see Fig. 7.1) and slightly modified the example movement detection code so that it counted and printed out the number of moves from frame to frame. The only changes made to the default OpenCV program were to add and print a counter to show the number of times movement occurred at each time frame (*MV_count*) in the main loop:

```
if self.frame_idx % self.detect_interval == 0:
        mask = np.zeros_like(frame_gray)
        mask[:] = 255
        for x, y in [np.int32(tr[-1]) for tr in self.tracks]:
            cv2.circle(mask, (x, y), 5, 0, -1)
        p = cv2.goodFeaturesToTrack(frame_gray, mask = mask,
        **feature_params)
        if p is not None:
            for x, y in np.float32(p).reshape(-1, 2):
                self.tracks.append([(x, y)])
                MV_count = MV_count+1;
```

```
        self.frame_idx += 1
        self.prev_gray = frame_gray
        cv2.imshow('lk_track', vis)

    if MV_count > 0:
        print( MV_count )
        MV_count = 0;

    ch = cv2.waitKey(1)
    if ch == 27:
        break
```

The Lucas–Kanade sparse optical flow demo uses goodFeaturesToTrack for track initialization and back-tracking for match verification between frames. The default draw-rate of the lk_track.py example code is every 5 frames of video; we modified this to 15 frames to reduce the granularity of the information printed out. Because the loop only redraws dots that have changed, their count can be taken as a measure of overall movement in the image. Because the cameras and background are static, the only movement can come from the facilitator (except for the very rare frames when a subject moves into view of the back camera). A batch shell script called this program for every video in the corpus and wrote the results to a text file which we read into R (R Core Team 2012) for analysis and plotting.

Detailed local information, such as a nodding of the head, twiddling of thumbs, heaving of the chest, and full-body heave (sigh or yawn) is deliberately omitted from this study, though there are tools available for their tracking and analysis. Plenty of examples can be found in the corpus recordings.

Figure 7.2 shows a sample plot of output from the movement counting. The spikes show a larger number of tracked points to have changed relative to the previously tracked frame. These spikes represent hand, face, or body movement from the facilitator, as registered by Camera3. We decided that a simple measure of this movement could be determined by counting the differences from frame to frame, where little movement would result in a small number and considerable movement would produce a larger one. A certain amount of movement detection is to be expected from noise in the image processing itself. Since absolute values were not of interest to us at this stage, we simply took the ratio of 'little-activity' to 'more-activity' as our score for each conversation. Expressed in R, this becomes: $t = table(abs(diff(dat))) > 3$); $t[1]/t[2]$ which produces a single ratio value expressing the relative amount of activity from the facilitator for each conversation. These values were then compared for each of the categories of conversation in the corpus.

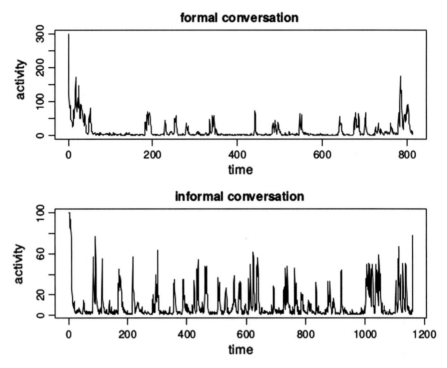

Fig. 7.2 Movement tracks for informal and formal conversations. The lower plot (longer in duration but compressed for display) shows more activity spikes for the informal conversation than in the formal case shown in the upper plot. In both cases, clear portions of inactivity (or very low levels of activity) are shown while the facilitator is monitoring/listening

7.5 Discussion of the Results

Each recording in the corpus can be categorised (from the filename codes) into male versus female subject, formal versus informal conversation, city versus country background of the subject, and age of subject (in years) as well as by interlocutor ID. Our analysis included 220 recordings, half formal and half informal; there were 118 male participants and 102 female. 134 of these were city people, while 86 were from village backgrounds.

Ages of the participants are as follows (count in brackets):
19(14) 20(50) 21(52) 22(22) 23(18) 24(22) 25(16) 26(10) 27(4) 28(4) 29(4) 30(4).

Figure 7.3 shows boxplots of the four main variables. We can see from the figure that the facilitator made little distinction in her degree of movement (or involvement in the conversation) between men and women, city and village people, and age of subject—the differences are not significant; but as expected, there was a significant difference in her behaviour between the formal and informal interview situations.

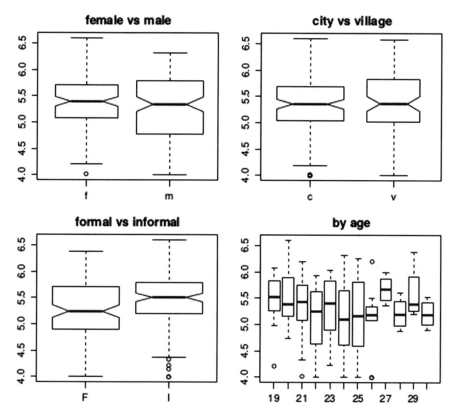

Fig. 7.3 Boxplots showing the measure of movement for each category of conversation. Here high indicates more movement—no significant difference is observed for any but the formal/informal condition. Differences can be considered significant if the notches do not overlap

A Welch Two Sample t-test gave t = 2.7019, df = 216.754, p-value = 0.00744 for the difference in formal and informal movement means.

Having thus established the veracity of the measure, we can now move on to the more interesting part of our study: the comparison between speaker ID and movement score. Figure 7.4 shows that the behaviour of the facilitator was consistently different for several interlocutors—i.e., she became more involved or had a higher degree of engagement in conversations with some people than with others.

An examination of the correlation of scores for each individual between the formal and informal conversations shows a consistent and different level of 'engagement' on the part of the facilitator for each: *cor* = 0.606. See Fig. 7.5 for a visual display of this correlation.

This is of course to be expected from a human being, given the nature of human society, but if true, this method can then be used by machines to estimate the degree of engagement of a human interlocutor in an automated dialogue with an autonomous agent.

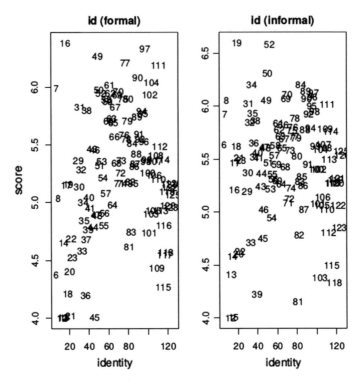

Fig. 7.4 Scores for each individual under the formal and informal conditions. There is considerable diversity in 'movement' but the ranking of the individuals is quite similar in each case. Activity score is plotted on the vertical axis, and speaker ID is ranked in numerical order along the horizontal axis. The same subject can be compared across conditions and several similarities in ranking can be observed

7.6   Conclusion

In this study we performed an image analysis of several conversations in two modalities with the goal of establishing a simple measure of engagement. We associated physical activity with involvement of the facilitator in each conversation and determined that there were differences in activity (and perhaps therefore engagement) according to certain prevailing conditions.

No significant differences in behaviour were found for conversations with male or female subjects. Nor were differences found between conversations with city-dwelling and village-based individuals There were differences in activity associated with the age of each subject, but no patterns or trends were observed that might indicate a higher-level explanation for this variability. However, as expected, significant differences in activity (and engagement?) were found between the formal and informal modes of conversation.

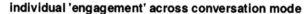

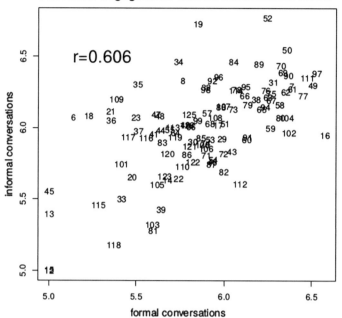

Fig. 7.5 A plot of the 'engagement' scores for each individual across formal and informal conversations shows a high correlation, confirming that the facilitator reacted consistently differently to the various individuals across conversations. We might infer from this that she became more involved with some individuals and less with others regardless of conversation mode

We then went further and interpolated an extension to the individual, on the grounds that some conversations might have been more interesting (and therefore more engaging) than others. Observation of the relevant videos confirmed that the facilitator did indeed behave differently towards different individuals, as would be expected, and that the ranking of the measure accorded with the difference in degree of observed engagement in the conversation.

We therefore propose the use of this or a similar measure derived from image processing to be used alongside the physical acoustic and prosodic parameters for use in future annotation of the corpus, and look forward to future work where similarities between acoustic-based and image-based measures might be found.

Acknowledgements The author wishes to acknowledge the support of AdaptCentre.ie and the School of Computer Science and Statistics in Dublin, and is particularly grateful to Laszlo Hunyadi and in particular the technical help of István Szekrényes for making the data available. The ADAPT Centre for Digital Content Technology is funded under the SFI Research Centres Programme (Grant 13/RC/2106) and is co-funded under the European Regional Development Fund.

References

Bradski G (2000) The OpenCV library. Dr Dobb's J Softw Tools

Campbell N (2009) Tools and resources for visualising conversational-speech interaction. Springer, Berlin, pp 176–188. https://doi.org/10.1007/978-3-642-04793-0_11

Haider F, Campbell N, Luz S (2016) Active speaker detection in human machine multiparty dialogue using visual prosody information. In: 2016 IEEE global conference on signal and information processing (GlobalSIP), pp 1207–1211. https://doi.org/10.1109/GlobalSIP.2016.7906033

Hunyadi L, Váradi T, Szekrényes I (2016) Language technology tools and resources for the analysis of multimodal communication. In: Proceedings of the LT4DH. University of Tübingen, Tübingen, pp 117–124

Kendon A (1990) Conducting interaction: patterns of behavior in focused encounters. Cambridge University Press, Cambridge

Pápay K, Szeghalmy S, Szekrényes I (2011) Hucomtech multimodal corpus annotation. Argumentum 7:330–347

R Core Team (2012) R: a language and environment for statistical computing. R Foundation for Statistical Computing, Vienna. http://www.R-project.org/

Szekrényes I (2014) Annotation and interpretation of prosodic data in the HuComTech corpus for multimodal user interfaces. J Multimodal User Interfaces 8:143–150

Chapter 8
Applying Neural Network Techniques for Topic Change Detection in the HuComTech Corpus

György Kovács and István Szekrényes

Abstract In the age of The Internet we are generating documents (both written and spoken) at an unprecedented rate. This rate of document creation—as well as the number of already existing documents—makes manual processing time-consuming and costly to the point of infeasibility. This is the reason why we are in need of automatic methods that are suitable for the processing of written as well as spoken documents. One crucial part of processing documents is partitioning said documents into different segments based on the topic being discussed. A self-evident application of this would be for example partitioning a news broadcast into different news stories. One of the first steps of doing so would be identifying the shifts in the topic frame-work, or in other words, finding the time-interval where the announcer is changing from one news story to the next. Naturally, as the transition between news stories are often accompanied by easily identifiable audio—(e.g. signal) and visual (e.g. change in graphics) cues, this would not be a particularly different task. However, in other cases the solution to this problem would be far less obvious. Here, we approach this task for the case of spoken dialogues (interviews). One particular difficulty of these dialogues is that the interlocutors often switch between languages. Because of this (and in the hope of contributing to the generality of our method) we carried out topic change detection in a content-free manner, focusing on speaker roles, and prosodic features. For the processing of said features we will employ neural networks, and will demonstrate that using the proper classifier combination methods this can lead to a detection performance that is competitive with that of the state-of-the-art.

G. Kovács
Research Institute for Linguistics of the Hungarian Academy of Sciences, Budapest, Hungary
e-mail: gykovacs@inf.u-szeged.hu

MTA SZTE Research Group on Artificial Intelligence, Szeged, Hungary

Embedded Internet Systems Lab, Luleå University of Technology, Luleå, Sweden

I. Szekrényes (✉)
Institute of Philosophy, University of Debrecen, Debrecen, Hungary
e-mail: szekrenyes.istvan@arts.unideb.hu

© Springer Nature Switzerland AG 2020
L. Hunyadi and I. Szekrényes (eds.), *The Temporal Structure of Multimodal Communication*, Intelligent Systems Reference Library 164,
https://doi.org/10.1007/978-3-030-22895-8_8

147

8.1 Introduction

In the task of topic change detection the goal is to automatically identify shifts in the topic framework; that is to say, identify where one part of the conversation ends and another begins (James 1995). Once these positions have been successfully identified, this information can be used to create a topic segmentation. Such segmentations can be useful in general applications like audio-visual information retrieval (Galukov 2012) and document summarisation (Angheluta et al. 2002), as well as more specific tasks like following the progress of a news story across multiple broadcasts (Purver 2011).

Here, we take the first steps towards a content-free topic segmentation method, by presenting an algorithm for the detection of topic change, based on neural network techniques. We train and evaluate this algorithm on the two-party conversations of the HuComTech corpus (Hunyadi et al. 2016), and compare the resulting F-scores with those reported in similar studies.

The structure of this chapter is as follows. In this section, we survey the topic segmentation literature. Then in Sect. 8.2 we briefly describe the HuComTech corpus, the features selected for the task along with the feature extraction methods, and the train/development/test partitioning we applied. Afterwards in Sect. 8.3 we describe the methods used for classification, and the error metrics we used to measure their performance. This is followed by the presentation and discussion of our results in Sect. 8.4. Lastly, in Sect. 8.5 we draw some conclusions and suggest possible directions for future research.

8.1.1 Related Work

Earlier publications examining the topical structure of the HuComTech corpus focused on the fine-grained structure of conversations, attempting to identify small shifts in the topic structure, and even differentiate between different degrees of topic change based on its motivatedness (Kovács et al. 2016; Kovács and Váradi 2017). The notion of a topic however is difficult to define (Purver 2011), making it difficult even for humans to reliably mark topic boundaries (Gruenstein et al. 2005), especially when the aim is to mark the fine-grained topical structure (Gruenstein et al. 2008). This problem however can be alleviated by sticking to coarse-grained topics (Galley et al. 2003), and informing the annotators of the agenda of the meeting (Banerjee and Rudnicky 2007) or dialogue. For these reasons, in this study we focus on identifying topic changes that correspond to the agenda items of the dialogues in the HuComTech corpus. We do so, however, in a content-free manner, without relying on human note-keeping (Banerjee and Rudnicky 2007), or the availability of a transcript (Sapru and Bourlard 2014; Sheikh et al. 2017).

In many of studies on topic structure discovery, the usual approach involves the segmentation of text (Chifu and Fournier 2016; Choi 2000; Hearst 1994; Kozima

1993; Reynar 1994; Sitbon and Bellot 2007). And while there are studies that utilise acoustic indicators as well, many of these studies still rely on the availability of transcripts and lexical information. In the study of Grosz and Sidner (1986), experts determined elements of discourse structure based on text alone or based on text and speech. The same categories (text alone or text and speech) were used in the experiments of Hirschberg and Nakatani (1996) on direction-giving monologues. Other researchers employed decision trees (Beeferman 1999; Passonneau and Litman 1997) or Hidden Markov Models (HMMs) (Shriberg et al. 2000; Tür et al. 2001) to combine prosodic and lexical features.

There are cases, however, where one cannot rely on the availability of transcripts or lexical features for creating a topic segmentation. One possible reason for this is privacy concerns: Multidisciplinary Medical Team Meetings (MDMTs) may be prime candidates for topic segmentation tasks (e.g. segmenting the meetings into individual patient case discussions), but their sensitive content makes it difficult to gather sufficient training data (Luz 2009). Another reason might be the particular domain, and language one needs to work in, as there are situations where there is no transcript available for the data, and the performance of an automatic recogniser is not good enough for segmentation (Malioutov et al. 2007). The above reasons motivated the appearance of content-free techniques for topic segmentation. Malioutov et al. for example predict topic changes by analysing the distribution of acoustic patterns extracted directly from the audio data (Malioutov et al. 2007). The method we used in our study is similar to that of Luz and Su, who use the "simplest form of content-free representation, namely: duration of talk spurts, silences and speech overlaps, optionally complemented with speaker role information" (Luz and Su 2010) for dialogue segmentation. Here, however we also used prosodic features, such as fundamental frequency, and speech rate.

8.2   Research Material

In this study, the HuComTech multimodal corpus (Hunyadi et al. 2016) was used for training and testing purposes. The corpus contains 50 hours of spontaneous speech with the participation of 111 native Hungarian speakers between the age of 19 and 30. The recordings can be divided into two subsets, namely formal dialogues of simulated job interviews and the subsequent informal conversations (both of them are performed by the very same participants in spontaneous way). The interviewer (called "agent" in the corpus) was responsible for keeping up the conversation by asking questions and giving feedback to the speakers. In the informal conversation, they also told their own stories for making the experimental situation more relaxed. It was needed because the interviewees asked to talk about personal topics (e.g. "Tell me about a negative experience you've recently had") to an unfamiliar person. The average duration of job interviews was 10 min while the informal conversation took about 20 min.

The interviews were face to face conversations in sitting position. Two fixed microphones were used to record the speaker's audio using 44100 Hz sampling rate and 16 bit quantization. Video recordings are also available in the corpus, but in the actual experiment, all features were based on the processed audio: sequences of turn taking (speaker change), annotation of topic change and prosody.

8.2.1 Annotation of Topic and Speaker Change

The annotation of topic change marks those utterances of interactions when the interviewer asks one of the previously designed questions turning the conversation into a new direction. This type of speech act can be categorized as *unmotivated* topic shift because the new topic is not necessarily connected semantically to the previous one (for instance, after the "negative experience" they asked to tell a joke) and they divide the conversation into well separable blocks. Moving from one topic to another is not entirely "natural" and neither is the partner's cooperation to it. The topic change would only happen due to the special nature of the given situation: by the interviewer having the authority to manage the interview process. Our assumption was that this type of topic shift would be more likely to affect the prosodic structure of the conversation compared to other types when the topic change is motivated (the new and the previous topic are in a topic-subtopic or other associative relation), rather than only demanded by the interview-scenario.

For training and testing purposes, labels of speaker information (whether the actual turn belongs to the interviewer or the interviewee) are also included in our data, and they are based on the manual annotation of turn taking available in the HuComTech corpus. There are four possible labels for every turn: no one speaks ("silent"), the interviewer speaks ("agent"), the interviewee speaks ("speaker"), the interviewer and the interviewee speak at the same time ("overlapping speech"). Of course, it was an effective starting point for our experiment that in every case when the turn does not have the label "agent" or "overlap", the topic change has zero probability, because the topic shift is always performed by the "agent". However, Fig. 8.1 shows that the turns of "agent" did not always involve topic change either. In informal conversations, the ratio of topic shift *turn takes* is under 20% (as mentioned above, the interviewer often became "storyteller"), and even in the job interviews, the role of interviewer was not limited to asking new questions. Therefore our classification task was to isolate special, topic-shifting turns of agent from those other utterances when they did not want to give a new direction to the conversation.

8.2.2 Prosodic Features

In a former experiment (Szekrényes and Kovács 2017) to classify dialogue segments into formal and informal ones in the HuComTech corpus, we used ProsoTool

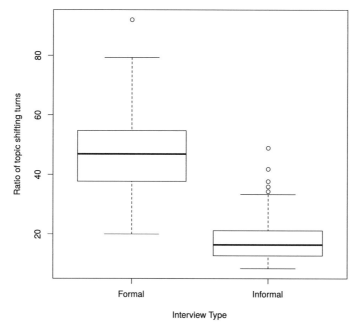

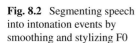

Fig. 8.1   The ratio of topic shifting turn takes in formal and informal interviews

Fig. 8.2   Segmenting speech into intonation events by smoothing and stylizing F0

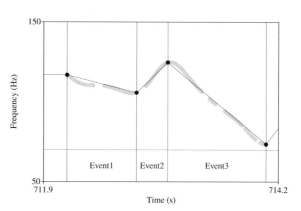

(Szekrényes 2015) for automatic labelling of intonation. The algorithm was implemented as a Praat script (Boersma 2016) and it was based on the stylization of the smoothed F0 contour. As can be seen in Fig. 8.2, the resulting intonation events follow the major turning points of the trend lines derived from F0. In the original version, these events are labeled using five categorical labels ("level", "ascending", "rise", "descending" and "fall") which describe the type of pitch modulation. The output was also extended with the relative position of breaking points in the individual vocal range of the speakers.

In this study, we applied the very same methodology for segmentation but the categorical labels of intonation events were replaced with continuous variables using two methods of normalization computed from (1) the F0 values associated with the breaking points (expressing "position"), (2) the F0 difference of the starting and ending points of a particular segment (expressing F0 trend), (3) the duration of the intonation events and the turns of speaker change.

In case of duration (as used for the turn-based and the segment-based classifiers, see details in Sect. 8.3) as well as the breaking points of intonation trends, zero was used to express the average value:

$$z_i = \frac{x_i - \bar{x}}{\sigma} \tag{8.1}$$

In case of intonation trends, zero denoted that there was no F0 modulation in the particular segment:

$$z_i = \frac{x_i - \bar{x}}{\left(\sigma - \frac{0 - \bar{x}}{\sigma}\right)} \tag{8.2}$$

In addition to pitch movement, two other prosodic features were used for detecting topic change. We measured intensity and speech rate in every intonation event applying the same normalization method as used for duration and breaking points of F0 trends. The speech rate was also calculated for the full turns of speaker change based on the algorithm of de Jong and Wempe (2009).

8.2.3 Train Development Test Partitioning

It is a good idea to have separate sets for training our machine learning models, for tuning their meta-parameters, and for evaluating these models fairly. In our experiments on the HuComTech corpus, we followed the 75/10/15 ratio defined by Kovács et al. for topical unit classification (Kovács et al. 2016).

8.3 Experimental Settings

The main modules utilised in the different methods for the classification of topic change are depicted in Fig. 8.3. As can be seen at the bottom of the figure, we experimented with three settings. In the baseline system, we investigated the performance that can be achieved when (similar to Luz and Su 2010) we just use turn-based speaker-role information combined with the length of each turn.

We also examined the classification performance attainable when we combine the speaker-role information with prosodic features such as the fundamental frequency

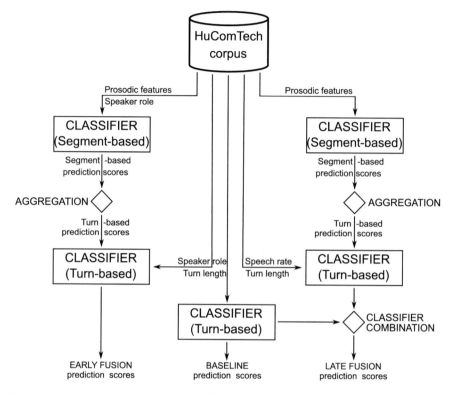

Fig. 8.3 Overview of the topic change classification methods applied here

and speech rate. One method of combination we investigated was where the two feature sets were simply merged before processing. We will refer to this combination method as "early fusion". When using this method we first get a prediction for each segment of dialogue based on the merged feature set of speaker-role information and prosodic features. Next, we merge the posterior probability estimates of every segment in a turn, and append the turn-length and speaker-role to these estimates. We will use the resulting feature set in a second classifier to estimate the turn-based probability of topic change.

The second method of combination we examined—and shall refer to it as "late fusion"—was built on the initial separation of prosodic features and speaker-role information. Here, we first calculate posterior estimates for each segment by just using the prosodic features. Then, we aggregate these estimates for each turn, and supplement this information with turn-based speech rate information, as well as turn-length information. We use the resulting feature vectors in a second classifier providing turn-based posterior estimates for topic change. Lastly, we combine the probability estimates of this model with those got using the baseline classifier to get our prediction scores for the late fusion combination method.

Here, we present a general outline of the three methods that will be investigated in our experiments. Below we will examine aspects of these methods that have not yet been discussed. These aspects include the classifiers we applied, the method used for the aggregation of segment-based prediction scores, the method used for classifier combination, and the metrics applied to compare the output performance of the three models examined here.

8.3.1 Probabilistic Sampling

One big problem a classifier needs to handle when working on the task of topic change classification is that of data imbalance. In a dialogue, instances of topic change are naturally much rarer than instances when the topic does not change. This imbalance in class distribution may cause a classifier to be biased towards the more common class, resulting in a worse classification performance of the rarer class (Lawrence et al. 1998). One way of overcoming this difficulty is to downsample the more common class, omitting some of these examples from the train set. This, however, would lead to the loss of important training data (Domingos 2012). Another possibility is to include samples from the rarer class in the train set. Although creating new data may be unfeasible, we can simulate having more samples from the rarer class by reusing the same samples. This is precisely what we do when we apply the method of probabilistic sampling. Here, we first select a class at random, and then draw a random sample from the selected class (Tóth and Kocsor 2005). The first step of selecting a random class here can be viewed as sampling from a multinomial distribution, given that each class has a probability $P(c_i)$ (Grósz and Nagy 2014). That is,

$$P(c_i) = \lambda(1/N) + (1 - \lambda)Prior(c_i) \qquad (1 \leq i, j \leq N; \lambda \in [0, 1]), \qquad (8.3)$$

where N is the number of classes, and the parameter λ controls the uniformity of the distribution. In this case a value of $\lambda = 0$ leads to the original distribution, while a value of $\lambda = 1$ leads to a uniform distribution (hence the $\lambda = 1$ case is also referred to as "uniform class sampling" (Tóth and Kocsor 2005).

8.3.2 Neural Network Architectures

We investigated two neural network architectures for the classification of dialogue-segments, and dialogue turns.

8.3.2.1 Deep Rectifier Neural Nets (DRNs)

Deep Rectifier Neural Nets are feedforward neural networks with more than one hidden layer that contain neurons that utilise the rectifier activation function (see Eq. 8.4) instead of the standard sigmoid activation function.

$$rectifier(x) = max(0, x) \qquad (8.4)$$

This architecture mitigates the effect of vanishing gradients, while also leading to sparser networks, making it a popular tool in speech technology (Grósz et al. 2015; Maas et al. 2013; Tóth 2013) and in other areas (Glorot et al. 2011; Grósz and Nagy 2014; Kovács et al. 2016) as well. The neural nets employed here had three hidden layers, each containing 250 neurons, and an output layer containing two neurons. We trained our DRNs on the training set, and used the development set for learn-rate scheduling purposes and finding good meta-parameter values. Meta-parameters here included the parameter λ used in probabilistic sampling, and the number of neighbouring feature vectors available for the classifier.

8.3.2.2 Gated Recurrent Units (GRUs)

The network of Gated Recurrent Units is a special case of the Recurrent Neural Network (RNN) architecture, complemented with an update gate and a reset gate to control the effect of the current input and previous memory on the current activation and current state (Dey and Salemt 2017). Similar to the rectifier activation function, GRUs were partly proposed to alleviate the problem of vanishing gradients (Cho et al. 2014). The resulting architecture was successfully applied in speech technology (Khandelwal et al. 2016; Lu et al. 2016), and in other applications (Jozefowicz et al. 2015; Kuta et al. 2017) as well. In our experiments we implemented a sequence classification method using GRU networks with Tensorflow (Abadi et al. 2015). Each of these GRU networks had three layers, consisting of 250 units, and were trained using the Adam optimiser. Similar to the training of DRNs, we used the training set for parameter value optimisation, and the development set for learn-rate scheduling, and the optimisation of meta-parameters (such as the parameter λ of the probabilistic sampling, and the number of neighbouring feature vectors used).

8.3.3 Classifier Combination

To combine the posterior probability estimates obtained from classifiers trained on different features (e.g. classifiers trained on prosodic features, and classifiers trained on speaker-role information), or different classifiers trained on the same set of features (e.g. GRU networks and DRNs trained on speaker-role information), we examined several different techniques. After our initial experiments we chose to combine our models by applying the 3 product rule (Dombi 2013) on their posterior probability estimates for the topic change class. This means that if y_i is the posterior probability estimate for the topic change class of the i-th classifier, we can get the combined estimate y for the probability of topic change using the following formula:

$$y = \frac{\prod_{i=1}^{n} y_i}{\prod_{i=1}^{n} y_i + \prod_{i=1}^{n} (1 - y_i)}, \tag{8.5}$$

where n is the number of classifiers used.

8.3.4 Error Metrics

There are several metrics available for measuring the performance of classifiers on tasks such as topic change detection. Here, we used two of these metrics, namely

- **Unweighted Average Recall (UAR)**: the imbalanced class distribution affects evaluation as well. Because of this, during training, when evaluating the performance of classifiers on the development set for the learning rate scheduler, we used the unweighted average recall metric. This measure is especially popular for the evaluation of tasks with an unbalanced class distribution (partly due to its usage at Interspeech challenges) (Rosenberg 2012). The UAR metric can be calculated using the confusion matrix A like so:

$$UAR = \frac{1}{N} \sum_{j=1}^{N} \frac{A_{jj}}{\sum_{i=1}^{N} A_{ij}}, \tag{8.6}$$

  where N is the number of classes, and A_{ij} is the number of instances in class j that are classified as instances of class i.
- **F score**: another possible way to evaluate our models on binary classification tasks (such as topic change classification) is to use the F_1-score, where we take the harmonic mean of precision and recall. That is,

$$F_1 = 2 \cdot \frac{precision \cdot recall}{precision + recall} \tag{8.7}$$

8.4 Results and Discussion

Here, we discuss the topic change classification results that we got in our experiments. First, we examine the results obtained using individual models. Then, we examine the results we obtained with classifier combination. Lastly, we compare our results with those reported in similar studies on topic change detection and topic segmentation.

Table 8.1 F-scores on the topic change detection task using individual models. The F-scores reported here represent the average performance of five independently trained models. The highest F-score is shown in bold

Method	Network model	Parameter λ	No. of. neighbours	F-score	
				Dev.	Test
Baseline	DRN	0.4	7	0.510	0.407
Baseline	GRU	0.1	5	0.501	0.423
Early fusion	DRN	0.3	11	**0.535**	0.437
Early fusion	GRU	0.1	5	0.523	**0.448**

8.4.1 Experiments with Individual Models

To ensure the comparability of results, we will first examine only those results that were obtained without the use of classifier combination. This means that Table 8.1 just contains the results got using the baseline model, and the results got using early fusion.

When comparing the results obtained using the baseline model with those obtained using early fusion, we see in Table 8.1 that regardless of the architecture (DRN or GRU) applied, we get better scores on both the development and the test set when prosodic features are included in the process. Furthermore, even the worse performing early fusion model attains better scores than those attained by the best baseline model.

When comparing the performance of different architectures, the results are not as easy to interpret. In the baseline model we find that while the F-scores obtained with the DRN networks are on average slightly higher on the development set than those obtained with the GRU networks, the opposite is true when we examine our results got on the test set. The same observation can be made with the results attained by the early fusion: the architecture that performs better on the development set just underperforms on the test set.

8.4.2 Experiments with Classifier Combination

Next, we compared the results obtained using late fusion with those obtained when applying classifier combination in the early fusion and baseline models. The results of these experiments are listed in Table 8.2. Here, the reported results represent the average of twenty-five results, as when using classifier combination, we applied our combination method to every possible pairing of the individual results. Furthermore, for each method we present two combinational results. First, the one got when combining the best individual models (i.e. the models that on average provided the best performance on the development set), then the one obtained when combining the

Table 8.2 F-scores on the topic change detection task using classifier combination. The F-scores reported here represent the average performance of twenty-five independently trained models. The highest F-score is shown in bold

Method	Network model	Feature set	Parameter λ	No. of. neigh.	F-score	
					Dev.	Test
Baseline	DRN	Speaker-role	0.4	7	0.519	0.432
	GRU	Speaker-role	0.1	5		
Baseline	DRN	Speaker-role	0.3	11	0.544	0.458
	GRU	Speaker-role	0.2	7		
Early fusion	DRN	Speaker-role + Prosodic	0.3	11	0.561	0.462
	GRU	Speaker-role + Prosodic	0.1	5		
Early fusion	DRN	Speaker-role + prosodic	0.3	11	0.561	0.462
	GRU	Speaker-role + Prosodic	0.1	5		
Late fusion	GRU	Speaker-role	0.1	5	0.549	0.427
	GRU	Prosodic	0.1	7		
Late fusion	GRU	Speaker-role	0.7	19	**0.597**	**0.490**
	GRU	Prosodic	0.0	7		

models that attain the best performance in combination (i.e. the models that when combined provided the best performance on the development set).

Table 8.2 makes clear that when the best individual models and the models that perform best in combination are different, the latter performs better on both the development set and the test set. We can also see in Table 8.2 that both early fusion and late fusion achieve higher F-scores than those got using the baseline model. And while according to the test of significance (a two-sampled t-test with unequal variance) the difference between the results got using the baseline, and the results got using the early fusion on the test set are not significant; the difference in the performance of the late baseline and late fusion methods is significant ($p < 9 \cdot 10^{-9}$). Lastly, it should be mentioned that in this case the model that performed best on the development set also proved to be the best performing model on the test set.

8.4.3 Comparison with the Results of Similar Studies

Although we are unaware of any study that attempts to handle the task on the same database, it may still be informative to compare our results with those reported in similar studies. Perhaps the study most similar to ours is that of Luz and Su (2010), who use content-free methods for the topic segmentation of an MDTM corpus (Kane and Luz 2006). Another study similar to ours that uses audio and prosodic features is by Hirschberg and Nakatani (1998), who used Classification and Regression Trees (CARTs) (Breiman et al. 1984) in the segmentation of the Boston Directions Cor-

Table 8.3 Topic change detection (the best results are shown in bold)

Method	Precision	Recall	F-score
Luz and Su (2010)	0.28	0.28	0.28
Molugu (2003)	0.20	**0.85**	0.33
Hirschberg and Nakatani (1998)	0.41	0.41	0.41
Baseline (current study)	0.46	0.46	0.46
Passoneau and Litman (1997)	0.47	0.50	0.48
Late fusion (current study)	**0.51**	0.48	**0.49**

pus. We also compared our results with those achieved using a combination of audio and text features. Passoneau and Litman (1997), for example, applied the C4.5 program (Quinlan 1993) for the combination of prosodic features, cue phrase features and noun phrase features. Here, we compare our results with those they reported on the held-back test set of their corpus. And Molugu reported his topic segmentation results using a combination of language modelling and prosodic modelling on the Hub4 news corpus (Molugu 2003).

As can be seen in Table 8.3, our baseline method using speaker-role information markedly outperforms the similar method of Luz and Su (2010) as well as the model of Hirschberg and Nakatani (1998). We also observe that among models whose results are presented here, the recall results produced by the model of Molugu were the highest, but due to the low precision scores, our baseline model outperformed this model as well, at least in terms of the F-score.

Table 8.3 also shows that while the F-score reported by Passoneau and Litman is higher than that produced by our baseline method, our late fusion model just outperforms theirs, even though it does not make use of lexical information, such as cue phrase features or noun phrase features.

Lastly, it should be mentioned that when working on broadcast news databases, much better F-score performances were attained. Amaral and Trancoso reported an F-score of 0.74 on the European Portuguese Broadcast News Corpus (Amaral and Trancoso 2009) using CARTs, while Calhoun attained an F-score of 0.79 using the same method on the Boston University Radio Corpus (Calhoun 2002). The two tasks are significantly different (Sherman and Liu 2008), making a comparison of results problematic.

8.5 Conclusions and Future Work

In this study we have made the first steps towards the content-free automatic topic segmentation of dialogues in the HuComTech corpus by introducing two methods for the classification of topic change turns. We found that the late fusion of prosodic and speaker-role information performs significantly better than the early fusion of the same sources of information, or the baseline method that relies solely on speaker-role information. We also compared our best result with other results taken from the topic change detection literature, and found that our method achieves higher F-scores than other content-free methods, and even methods that combine audio and lexical information. However, we should add that in many cases researchers do not report F-scores, and so a comparison of results was not always possible.

One possibility for the future is to examine more neural net models, such as networks of Long-Short Term Memory (LSTM) units, as well as bi-directional GRU networks. It might also be beneficial to reexamine current training processes, either by investigating more learning rates, or fine-tuning the process of probabilistic sampling. Another way to improve the classification scores would be to include other modalities in the process, such as video information. In the future we would also like to move from topic change classification to topic segmentation. For this, many methods may be investigated, such as simple filtering and Hidden Markov Models.

Acknowledgements The research reported in the paper was conducted with the support of the Hungarian Scientific Research Fund (OTKA) grant #K116938 and #K116402. Ministry of Human Capacities, Hungary grant 20391-3/2018/FEKUSTRAT is acknowledged.

References

Abadi M, Agarwal A, Barham P, Brevdo E, Chen Z, Citro C, Corrado GS, Davis A, Dean J, Devin M, Ghemawat S, Goodfellow I, Harp A, Irving G, Isard M, Jia Y, Jozefowicz R, Kaiser L, Kudlur M, Levenberg J, Mané D, Monga R, Moore S, Murray D, Olah C, Schuster M, Shlens J, Steiner B, Sutskever I, Talwar K, Tucker P, Vanhoucke V, Vasudevan V, Viégas F, Vinyals O, Warden P, Wattenberg M, Wicke M, Yu Y, Zheng X (2015) TensorFlow: large-scale machine learning on heterogeneous systems. https://www.tensorflow.org/, software available from tensorflow.org
Amaral R, Trancoso I (2009) Exploring the structure of broadcast news for topic segmentation. In: Vetulani Z, Uszkoreit H (eds) Human language technology. Challenges of the information society. Springer, Berlin, pp 1–12
Angheluta R, Busser RD, Moens MF (2002) The use of topic segmentation for automatic summarization. In: Workshop on text summarization in conjunction with the ACL 2002 and including the DARPA/NIST sponsored DUC 2002 meeting on text summarization, pp 11–12
Banerjee S, Rudnicky AI (2007) Segmenting meetings into agenda items by extracting implicit supervision from human note-taking. In: Proceedings of IUI, pp 151–159
Beeferman D, Berger JLA (1999) Statistical models for text segmentation. Mach Learn 34(1–3):177–210
Boersma DP, Weenink (2016) Praat: doing phonetics by computer [computer program]. version 6.0.22. http://www.praat.org/. Accessed 15 Nov 2016
Breiman L, Friedman J, Olshen R, Stone C (1984) Classification and regression trees. Taylor & Francis, London
Calhoun S (2002) Using prosody in ASR: the segmentation of broadcast radio news. Master's thesis, University of Edinburgh
Chifu AG, Fournier S (2016) SegChain: towards a generic automatic video segmentation framework, based on lexical chains of audio transcriptions. In: Proceedings of the 6th international conference on web intelligence, mining and semantics, pp 1–8
Cho K, van Merriënboer B, Gülçehre Ç, Bahdanau D, Bougares F, Schwenk H, Bengio Y (2014) Learning phrase representations using RNN encoder–decoder for statistical machine translation. In: Proceedings of the 2014 conference on empirical methods in natural language processing

(EMNLP), Association for Computational Linguistics, Doha, Qatar, pp 1724–1734. http://www.aclweb.org/anthology/D14-1179

Choi FYY (2000) Advances in domain independent linear text segmentation. In: Proceedings of NAACL, pp 26–33

de Jong NH, Wempe T (2009) Praat script to detect syllable nuclei and measure speech rate automatically. Behav Res Methods 41(2):385–390. https://doi.org/10.3758/BRM.41.2.385

Dey R, Salemt FM (2017) Gate-variants of gated recurrent unit (GRU) neural networks. In: 2017 IEEE 60th international midwest symposium on circuits and systems (MWSCAS), pp 1597–1600

Dombi J (2013) On a certain class of aggregative operators. Inf Sci 245:313–328

Domingos P (2012) A few useful things to know about machine learning. Commun ACM 55(10):78–87

Galley M, McKeown K, Fosler-Lussier E, Jing H (2003) Discourse segmentation of multi-party conversation. In: Proceedings of ACL, pp 562–569

Galukov P (2012) Application of topic segmentation in audiovisual information retrieval. In: Proceedings of WDS, pp 118–122

Glorot X, Bordes A, Bengio Y (2011) Deep sparse rectifier neural networks. In: Proceedings of AISTATS, pp 315–323

Grosz BJ, Sidner CL (1986) Attention, intentions, and the structure of discourse. Comput Linguist 12(3):175–204

Grósz T, Nagy I (2014) Document classification with deep rectifier neural networks and probabilistic sampling. In: Proceedings of TSD, pp 108–115

Grósz T, Busa-Fekete R, Gosztolya G, Tóth L (2015) Assessing the degree of nativeness and Parkinson's condition using Gaussian processes and deep rectifier neural networks. In: Proceedings of Interspeech, pp 1339–1343

Gruenstein A, Niekrasz J, Purver M (2005) Meeting structure annotation: data and tools. In: Proceedings of SIGDIAL, pp 117–127

Gruenstein A, Niekrasz J, Purver M (2008) Meeting structure annotation. In: Dybkjær L, Minker W (eds) Recent trends in discourse and dialogue. Springer, Netherlands, pp 247–274

Hearst MA (1994) Multi-paragraph segmentation of expository text. In: Proceedings of the ACL, pp 9–16

Hirschberg J, Nakatani CH (1996) A prosodic analysis of discourse segments in direction-giving monologues. In: Proceedings of the ACL, pp 286–293

Hirschberg J, Nakatani CH (1998) Acoustic indicators of topic segmentation. In: Proceedings of ICSLP

Hunyadi L, Váradi T, Szekrényes I (2016) Language technology tools and resources for the analysis of multimodal communication. In: Proceedings of LT4DH, University of Tübingen, Tübingen, pp 117–124

James AD (1995) Topic shift in casual conversation. Totem: Univ West Ont J Anthropol 2(1)

Jozefowicz R, Zaremba W, Sutskever I (2015) An empirical exploration of recurrent network architectures. In: Proceedings of ICML, pp 2342–2350

Kane B, Luz S (2006) Multidisciplinary medical team meetings: an analysis of collaborative working with special attention to timing and teleconferencing. Comput Support Coop Work 15(5–6):501–535

Khandelwal S, Lecouteux B, Besacier L (2016) Comparing GRU and LSTM for automatic speech recognition. Research report, LIG. https://hal.archives-ouvertes.fr/hal-01633254

Kovács G, Váradi T (2017) Examining the contribution of various modalities to topical unit classification on the HuComTech corpus (in Hungarian). In: Proceedings of MSZNY, pp 193–204

Kovács G, Grósz T, Váradi T (2016) Topical unit classification using deep neural nets and probabilistic sampling. In: Proceedings of CogInfoCom, pp 199–204

Kozima H (1993) Text segmentation based on similarity between words. In: Proceedings of the ACL, pp 286–288

Kuta M, Morawiec M, Kitowski J (2017) Sentiment analysis with tree-structured gated recurrent units. In: Proceedings of TSD, pp 74–82

Lawrence S, Burns I, Back A, Tsoi AC, Giles CL (1998) Neural network classification and prior class probabilities. In: Orr GB, Müller KR (eds) Neural networks: tricks of the trade. Springer, Berlin, pp 299–313

Lu L, Zhang X, Renais S (2016) On training the recurrent neural network encoder-decoder for large vocabulary end-to-end speech recognition. In: Proceedings of ICASSP, pp 5060–5064

Luz S (2009) Locating case discussion segments in recorded medical team meetings. In: Proceedings of the third workshop on searching spontaneous conversational speech, SSCS '09. ACM, pp 21–30

Luz S, Su J (2010) Assessing the effectiveness of conversational features for dialogue segmentation in medical team meetings and in the AMI corpus. In: Proceedings of SIGDIAL, pp 332–339

Maas AL, Hannun AY, Ng AY (2013) Rectifier nonlinearities improve neural network acoustic models. In: Proceedings of ICML, vol 30/1

Malioutov I, Park A, Barzilay R, Glass J (2007) Making sense of sound: unsupervised topic segmentation over acoustic input. In: Proceedings of the ACL, pp 504–511

Molugu MC (2003) Topic segmentation. Master's thesis, University of Edinburgh

Passonneau RJ, Litman DJ (1997) Discourse segmentation by human and automated means. Comput Linguist 23(1):103–139

Purver M (2011) Topic segmentation. In: Tur G, de Mori R (eds) Spoken language understanding: systems for extracting semantic information from speech. Wiley, New York, pp 291–317

Quinlan JR (1993) C4.5: programs for machine learning. Morgan Kaufmann Publishers Inc., San Francisco

Reynar JC (1994) An automatic method of finding topic boundaries. In: Proceedings of the ACL, pp 331–333

Rosenberg A (2012) Classifying skewed data: importance weighting to optimize average recall. In: Proceedings of Interspeech, pp 2242–2245

Sapru A, Bourlard H (2014) Detecting speaker roles and topic changes in multiparty conversations using latent topic models. In: Proceedings of Interspeech, pp 2882–2886

Sheikh I, Fohr D, Illina I (2017) Topic segmentation in ASR transcripts using bidirectional RNNs for change detection. In: Proceedings of ASRU

Sherman M, Liu Y (2008) Using hidden Markov models for topic segmentation of meeting transcripts. In: Proceedings of SLT, pp 185–188

Shriberg E, Stolcke A, Hakkani-Tür D, Tür G (2000) Prosody-based automatic segmentation of speech into sentences and topics. Speech Commun 32(1–2):127–154

Sitbon L, Bellot P (2007) Topic segmentation using weighted lexical links (WLL). In: Proceedings of the 30th annual international ACM SIGIR conference on research and development in information retrieval, SIGIR '07, pp 737–738

Szekrényes I (2015) Prosotool, a method for automatic annotation of fundamental frequency. In: 6th IEEE international conference on cognitive infocommunications (CogInfoCom). IEEE, New York, pp 291–296

Szekrényes I, Kovács G (2017) Classification of formal and informal dialogues based on turn-taking and intonation using deep neural networks. In: Karpov A, Potapova R, Mporas I (eds) Speech and computer. Springer International Publishing, Cham, pp 233–243

Tóth L (2013) Phone recognition with deep sparse rectifier neural networks. In: Proceedings of ICASSP, pp 6985–6989

Tóth L, Kocsor A (2005) Training HMM/ANN hybrid speech recognizers by probabilistic sampling. In: Proceedings of ICANN, pp 597–603

Tür G, Hakkani-Tür DZ, Stolcke A, Shriberg E (2001) Integrating prosodic and lexical cues for automatic topic segmentation. CoRR 31–57

Printed in the United States
By Bookmasters